Ministry of Agriculture, Fisheries and Food
The Scottish Office Agriculture and Fisheries Department
Department of Agriculture for Northern Ireland
Welsh Office Agriculture Department

The Digest of Agricultural Census Statistics

United Kingdom 1991

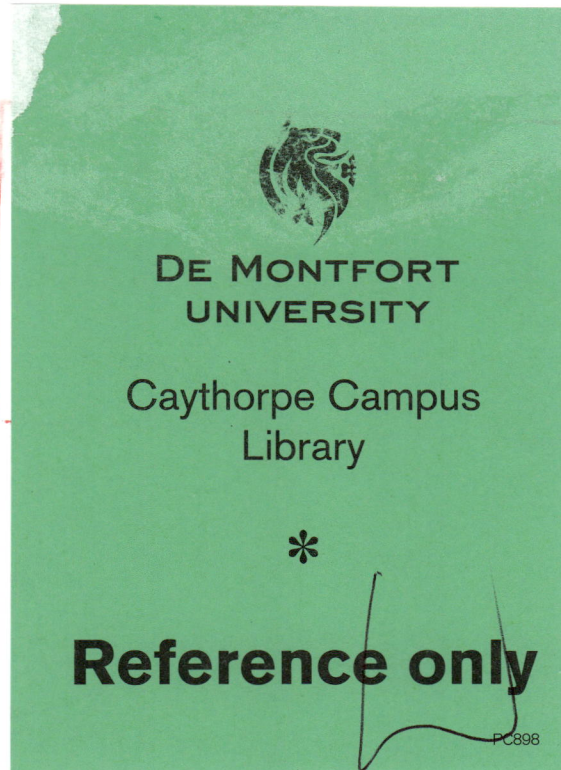

A publication of the Government Statistical Service

London : HMSO

Contents

Chapter 1
Introduction

The Digest and the Census

This is the first publication of the Digest of Agricultural Census Statistics. The Digest brings together the main results of the annual agricultural censuses which are held in parallel every June in England, Wales, Scotland and Northern Ireland. Aggregate figures are shown for each territory and for the United Kingdom as a whole for each year in the period 1982 to 1991. More detailed analyses are given for 1991 including summary data down to county level (regions in Scotland).

The census provides the basic physical statistics of farming: areas of land use and of crops; numbers of main livestock types ie cattle, sheep, pigs and poultry; and the numbers and kinds of persons working on the farm. All figures relate to the areas of land and numbers of animals and persons on the agricultural holding on the census date, which is the first weekday in June. See Chapter 2 for more information on the census.

The Shape of the Digest

Details of the tables are given in the Contents list on pages iv and v. The general pattern is as follows:

Chapter 3: Five- and ten-year comparisons for the UK; 1981 - 1986 - 1991

These tables show the main physical aggregates of land, crops, cattle, sheep, pigs, poultry and labour for the United Kingdom for 1981, 1986 and 1991, together with percentage changes over each of the two five-year periods. This provides a rapid reference to medium term movements; inspection of the annual tables in Chapter 4 is necessary for a more detailed appraisal of trends.

Chapter 4: Annual reference tables; 1982 to 1991

These tables comprise the heart of the Digest, with aggregates for all items of significance for each of the ten years 1982 to 1991:

Land Use and Crops		
Cattle and Calves		.... United Kingdom
Pigs		.... England
Sheep	**... for each of ....**	.... Wales
Poultry		.... Scotland
Labour		.... Northern Ireland

Chapter 5: Regional and county data for 1991

Summary information is provided for each of the eleven Standard Statistical Regions (SSRs) of the United Kingdom. The territories of Wales, Scotland and Northern Ireland each comprise one SSR; the other eight SSRs lie in England. At sub-SSR level, the same information is given for each English, Welsh and Northern Ireland county and for each equivalent area in Scotland (where such areas are termed "regions").

This summary information includes not only the key figures on land use, livestock and labour but also gives simple distributions of farms by size and by type. Percentage figures are also provided in order to show the relative importance of the different items.

Chapter 6: Maps of selected main items; 1991

The sixteen maps illustrate the spatial distribution of some of the key items. See the introduction to Chapter 6 for guidance on interpretation.

Chapter 7: Frequency distributions of main holdings by size; 1991

These tables show the size pattern of holdings, for England, Wales, Scotland, Northern Ireland, and for the United Kingdom as a whole, according to different main items. For each size band, both the number of holdings and the total of the item are shown, together with percentages, to permit simple inspection of concentration patterns.

Statistical Publications

It is intended that this Digest should be published annually, to include the latest year's final census figures and to present the latest ten years of main aggregate data together for ease of reference. The final results and analyses of each June census are available early in the following year. See the description of census operations on pages 2-3 and 2-4. In a normal year with a settled production cycle, the Digest should be published about the middle of the year following the census.

For agricultural census data, this Digest is the successor to the previous annual publication "Agricultural Statistics, United Kingdom" and thus continues a long succession of agricultural census publications from 1866. Other material included in the previous "Agricultural Statistics" volume such as agricultural prices and production is now published elsewhere.

The main related official publications containing agricultural statistics are as follows:

Agriculture in the United Kingdom; MAFF; annual
Farm Incomes in the United Kingdom; MAFF; annual
Basic Horticultural Statistics for the United Kingdom; MAFF; annual
Agricultural Market Report; MAFF; weekly prices
Agricultural Prices Indices; MAFF; monthly
Economic Report on Scottish Agriculture; Scottish Office; annual
Welsh Agricultural Statistics; Welsh Office; annual
Statistical Review of Northern Ireland Agriculture; Department of Agriculture for Northern Ireland; annual

Detailed tabulations of the agricultural censuses are available, in both standard form and by special analysis, on appropriate payment. The Agricultural Departments also publish the results of a range of sample surveys of farms drawn from the census. Enquiry may be made to the following offices:

For England and the United Kingdom:
Ministry of Agriculture, Fisheries and Food,
Agricultural Census Branch F,
Room A615, Government Buildings,
Epsom Road,
Guildford,
SURREY GU1 2LD
Telephone: Guildford (0483) 403577/403520

For Wales:
Welsh Office Agriculture Department,
Economic and Statistical Services Division,
Crown Buildings,
CARDIFF CF1 3NQ
Telephone: Cardiff (0222) 825052

For Scotland:
Scottish Office Agriculture and Fisheries Department,
Economics and Statistics Unit,
Pentland House,
47 Robbs Loan,
EDINBURGH EH14 1TW
Telephone: Edinburgh (031) 556 8400 ext. 6149

For Northern Ireland:
   Department of Agriculture for Northern Ireland,
   Dundonald House,
   Upper Newtownards Road,
   BELFAST BT4 3TB
   Telephone: Belfast (0232) 650111 ext. 528

Users of agricultural census statistics may also wish to know that the European Community Farm Structure Survey is conducted by Member States every two or three years. This survey provides a wide range of census-type data on broadly consistent definitions for each of the twelve countries of the Community. For the United Kingdom, the annual census is the source for most of the Farm Structure Survey data, but is supplemented by a special sample covering labour input ie work done in person-years, rather than the number of persons on farms on census day.

Results are available from the EC Farm Structure Surveys for 1975, 1979/80, 1983, 1985 and 1987. The latest survey was centred around 1990 and data from this are being processed. Further surveys are due in 1993, 1995 and 1997. Enquiries about statistics from the EC Farm Structure Survey may be made either to:

   Ministry of Agriculture, Fisheries and Food,
   Agricultural Census Branch E,
   Room A607, Government Buildings,
   Epsom Road,
   Guildford,
   SURREY GU1 2LD
   Telephone: Guildford (0483) 403728

or direct to:

   Office of Official Publications of the European Community,
   2 rue Mercier,
   L - 2985 LUXEMBOURG
   Telephone: Luxembourg (010 352) 499281

Chapter 2
Definitions and Operations

The agricultural holding

The basic unit of enumeration in the agricultural census is the holding or, more loosely, the "farm". The guideline definition is pragmatic and operational and subject to agreement with the individual farmer. The holding comprises land on which agricultural activities are carried out and which is by and large farmed in one unit having regard to such supplies as machinery, livestock, feedingstuffs and manpower, and to the distance between any separate areas of land involved and their type of farming. "Farming" includes horticultural activity.

The farmer is the "occupier" of the holding, the land of which may be owned or rented in whole or in part. The farm business may be operated by a "farmer" who is a single individual, a partnership of individuals, a limited company, or an institution of some kind. Some farmers occupy more than one holding.

Main and minor holdings

The annual census encompasses the 241 000 main agricultural holdings (1991) in the United Kingdom. One of the main purposes of the census is to estimate reliably the aggregates of the individual items collected. To this end, it is not necessary to include in the annual census the "minor" holdings which are below a certain size as they contribute only a small proportion of the totals. The annual census therefore comprises the main holdings above the specified size threshold.

The threshold which defines the operational difference between main and minor holdings has changed over time and varies somewhat between England and Wales, Scotland, and Northern Ireland. Periodic assessments are made of the number and total activity of minor holdings. The current position in each territory is as follows:-

England and Wales

A holding is classified as minor if all the following criteria are true:
   a. the total area is less than 6 hectares
   b. there is no regular whole-time farmer or worker
   c. the estimated annual labour requirement is less than 100 days (of 8 hours productive work by an adult worker under average conditions)
   d. the glasshouse area is less than 100 square metres
   e. the occupier does not farm another holding.

Note that if any of these conditions is not satisfied the holding is categorised as "main" and is sent a census form in June.

England and Wales (cont'd)

A simplified census of minor holdings is taken periodically. Transfers between main and minor are normally made only after each such census of minor holdings.

Note that aggregate estimates for minor holdings in England and Wales are included in the aggregates given in the tables of Chapters 3 and 4, but **not** in the regional and county data of Chapters 5 and 6 or in the size distributions of Chapter 7.

Such estimates for minor holdings are based on the latest Minor Holdings census for England and Wales, held in 1989. Following this, 11 200 main holdings in England and 1 900 in Wales were reclassified as minor. 3 200 previously minor holdings in England and 700 in Wales were reclassified as main holdings. Since the previous Minor Holdings Census of 1983, some 10 600 minor holdings in England were estimated to have ceased agricultural activity.

The table below shows the broad estimated aggregates for minor holdings in England and Wales after the reclassifications noted above, together with the 1991 aggregates for main holdings, and the minors as a percentage of the totals for all holdings.

MINOR HOLDINGS: Estimated Aggregates and Relative Contribution; 1991

	England			Wales			England and Wales		
	Minor holdings 1989	All holdings (1)	Minors as per cent of all	Minor holdings 1989	All holdings (1)	Minors as per cent of all	Minor holdings 1989	All holdings (1)	Minors as per cent of all
Number of holdings	37.7	188.6	19.9%	6.8	36.5	18.7%	44.5	225.2	19.7%
	thousand hectares			thousand hectares			thousand hectares		
Total area	88.5	9 420.5	0.9%	18.7	1 511.0	1.2%	107.2	10 931.5	0.9%
Crops and fallow	7.2	4 254.4	0.1%	0.1	71.8	0.2%	7.4	4 326.3	0.1%
Grass	60.1	3 944.6	1.5%	14.3	1 052.0	1.3%	74.4	4 996.7	1.4%
Rough grazing	16.1	742.5	2.1%	3.4	335.4	1.0%	19.5	1 078.0	1.8%
Woods and other land (2)	4.8	478.8	1.0%	0.8	51.5	1.5%	5.6	530.4	1.0%
	thousands			thousands			thousands		
Cattle	50.0	6 881.5	0.7%	8.0	1 343.3	0.6%	58.1	8 224.9	0.7%
Sheep	189.4	20 438.7	0.9%	68.6	10 850.8	0.6%	258.1	31 289.5	0.8%
Pigs	16.6	6 411.7	0.2%	1.0	103.1	1.0%	17.7	6 514.8	0.2%
Poultry (incl turkeys)	310.3	107 823.8	0.2%	37.2	7 246.4	0.5%	347.6	115 070.3	0.3%

Notes:

(1) main holdings 1991 plus minor holdings 1989

(2) including Set Aside land

Scotland

In Scotland there were about 20 000 minor holdings in 1991. A minor holding is defined as one with a low level of agricultural economic activity. Currently, one-third of minor holdings are surveyed each year in Scotland on a rotational basis.

Scotland has not included data in respect of their minor holdings in the tables appearing in this volume.

Northern Ireland

In Northern Ireland in 1991 there were about 15 000 minor holdings. These are holdings which satisfy all of the following criteria:

    a. the total area of land owned, or taken on long-term lease is less than 6 hectares.
    b. the business size is less than 1 British Size Unit which is defined as 2 000 European units of account of standard gross margin at average 1978-1980 values (the standard gross margin is defined on page 5-4).
    c. there is no regular full-time worker other than the owner.

One-third of minor holdings are surveyed each year in Northern Ireland on a rotational basis.

Northern Ireland has not included data in respect of their minor holdings in the tables appearing in this volume. In chapters 5, 6 and 7 the data also exclude low activity holdings.

Operations of the Agricultural Census

The following brief description of the operations of the agricultural census is based on the censuses of England and Wales. The censuses in Scotland and Northern Ireland follow similar procedures.

The central list of agricultural holdings is a key part of the census system. It is maintained and kept up-to-date by information which comes in broadly equal measure from the operation of the census itself and the many sample surveys based on it and from the Agriculture Departments' administrative dealings with farmers.

The farm list is used to send census forms to farmers and growers in mid-May, ready for the census date in early June. The census is held under the Agricultural Statistics Act 1979 (as amended in 1984). Occupiers have a legal obligation to make the census return, and the Act provides protection against disclosure or misuse of a farmer's information collected for statistical purposes, that is for aggregation with other returns.

The census is a postal enquiry and most farmers send back their completed form quickly, often also completing and keeping a "retention copy" for their own records. There is a sequence of reminders by postcard to prompt those who are a little late. The information on the forms is entered into the computer system and subjected to a battery of checks which examine consistency of entries, including those based on comparison with previous returns. Queries are raised with the farmer, often by telephone.

Within about six weeks of the census date enough forms have been received, and corrected if necessary, for the production of provisional results based on about 60 per cent of the whole census. These aggregates are examined and cleared by the census statisticians. At this stage, totals for the census are estimated by statistical methods, working from the information so far and "raising" the figures to reflect the full list of main holdings. Each Agricultural Department publishes the figures for its territory: the United Kingdom results are collated from these by the Ministry of Agriculture, Fisheries and Food and published, with the England provisional results, usually in the second half of August.

The census operations then continue towards maximising the response from farmers. After the census has been closed, any missing information is derived by statistical methods at the level of the holding, so that an estimated record is constructed for each holding in question. The full set of records for main holdings is used for the production of final results, including aggregates and a wide range of standard and specialised analyses.

The census is thus a major statistical exercise. Its outcome is subject to some margin of statistical error on account of the quality of the basic material and the methods which are used in adjusting and statistically treating the data for non-respondents. The census should not be regarded as a stocktaking exercise carried out to accounting standards and precision. It is debatable whether it has ever achieved such accuracy in its long history, but certainly today it must be seen as a large-scale data collection, executed to modern statistical standards.

Questions on methods and content of the agricultural census in each territory of the United Kingdom should be addressed to the appropriate Agricultural Department (see the contact points given in Chapter 1). Copies of blank census forms are available on request. The farmer's "retention copy" also includes notes for guidance for completing some of the items on the form.

The farm list and the annual census are the twin foundations of the system of statistical surveys of farms. With recent data for most holdings and the use of statistical methods it is possible to design sample surveys and specialised censuses of particular activities which are highly efficient for the estimation of a range of items. Examples are: livestock and autumn-sown crops in December; pig numbers in April and August; open-air horticulture in October and glasshouse in December; production of cereals, oilseeds, and peas and beans at the end of the harvest; stocks of cereals and amounts of grain fed on farm throughout the year; earnings and hours of workers monthly. Also, major specialised surveys are held from time to time, such as on irrigation practice, on the type and age of fruit trees, and on the amount of labour used on farms.

Conventional signs and rounding

..   means less than half the final digit shown

*   means data supressed to prevent disclosure of individual holdings

na   means not available

Totals may not necessarily agree with the sum of their components due to rounding.

Percentages are based upon unrounded figures.

Chapter 3

Five- and ten-year comparisons

United Kingdom

1981 - 1986 - 1991
with percentage changes

Five- and ten-year comparisons
Agricultural Land by TYPE of USE
United Kingdom (a)

Table 3.1
1981, 1986 and 1991 (at June Census)

thousand hectares

	1981	% change 1986 / 1981	1986	% change 1991 / 1986	1991
Total crops	**4 995.1**	**+4.9%**	**5 239.4**	**-5.4%**	**4 956.0**
+Bare fallow	76.1	-36.9%	48.0	+33.1%	63.9
=Total tillage	**5 071.2**	**+4.3%**	**5 287.4**	**-5.1%**	**5 019.8**
+Grasses under five years old	1 910.9	-9.8%	1 723.3	-8.3%	1 580.8
=Total arable land	**6 982.1**	**+0.4%**	**7 010.6**	**-5.8%**	**6 600.6**
+Grasses five years old and over	5 102.6	-0.5%	5 077.3	+3.7%	5 267.0
= Total tillage and grass	**12 084.6**	**+0.0%**	**12 087.9**	**-1.8%**	**11 867.6**
+Sole right rough grazing (b)	5 021.3	-3.8%	4 828.5	-3.2%	4 674.0
+Woodland on holdings	276.7	+14.3%	316.4	+16.3%	367.9
+All other land on holdings (c)	210.8	+7.6%	226.9	+51.6%	344.1
=TOTAL AREA ON HOLDINGS	**17 593.5**	**-0.8%**	**17 459.7**	**-1.2%**	**17 253.6**
Total crops	**4 995.1**	**+4.9%**	**5 239.4**	**-5.4%**	**4 956.0**
Total cereals (excluding maize)	**3 979.2**	**+1.1%**	**4 024.2**	**-13.0%**	**3 499.8**
Wheat	1 491.1	+34.0%	1 997.4	-0.8%	1 980.5
Barley - total	2 327.4	-17.7%	1 916.1	-27.3%	1 392.6
- winter	*na*	*na*	959.9	-12.4%	841.3
- spring	*na*	*na*	956.2	-42.3%	551.3
Oats	143.6	-32.4%	97.1	+6.5%	103.5
Mixed corn for threshing	10.6	-37.3%	6.7	-44.2%	3.7
Rye for threshing	6.5	+6.0%	6.9	+24.7%	8.6
Triticale	*na*	*na*	*na*	*na*	11.0
Total other arable crops not for stockfeeding	**539.3**	**+30.5%**	**704.0**	**+30.2%**	**916.4**
Potatoes (early and maincrop)	191.2	-7.1%	177.7	-0.6%	176.6
Sugar beet (not for stockfeeding)	210.2	-2.4%	205.0	-4.6%	195.7
Hops	5.7	-23.9%	4.4	-15.1%	3.7
Oilseed rape	125.0	+139.3%	299.1	+47.1%	439.9
Linseed	*na*	*na*	*na*	*na*	91.9
Other crops not for stockfeeding (d)	7.2	+146.2%	17.8	-51.1%	8.7
Total crops mainly for stockfeeding	**221.4**	**+34.4%**	**297.5**	**+13.0%**	**336.3**
Field beans	45.4	+31.5%	59.7	+119.2%	130.9
Peas for harvesting dry (e)	*na*	*na*	90.6	-20.7%	71.9
Other fodder crops (including maize for threshing)	176.0	-16.4%	147.2	-9.3%	133.5
Total horticultural crops	**255.2**	**-16.3%**	**213.6**	**-4.7%**	**203.5**
Vegetables for human consumption (f)	178.2	-18.1%	145.8	-4.6%	139.2
Orchards	*na*	*na*	37.9	-11.2%	33.6
Small fruit (g)	18.2	-16.2%	15.2	-3.9%	14.6
Hardy nursery stock, bulbs and flowers	12.7	-2.0%	12.4	+10.7%	13.8
Area under glass or plastic covered structures	2.2	+2.0%	2.3	+2.2%	2.3

Note

Please refer to notes at Table 4.1, pages 4-2 and 4-3.

Five- and ten-year comparisons
CATTLE and CALVES on Agricultural Holdings
United Kingdom (a)
1981, 1986 and 1991 (at June Census)

Table 3.2   thousands

	1981	% change 1986 1981	1986	% change 1991 1986	1991
TOTAL CATTLE AND CALVES	13 138.3	*-4.6%*	12 533.5	*-5.3%*	11 866.0
Total breeding herd	4 611.0	*-3.6%*	4 446.4	*-0.2%*	4 435.9
Dairy herd - total	3 190.9	*-1.7%*	3 138.1	*-11.7%*	2 769.7
cows and heifers in milk	2 906.7	*-1.3%*	2 868.8	*-14.9%*	2 440.0
cows in calf but not in milk	284.2	*-5.2%*	269.3	*+22.4%*	329.7
Beef herd - total	1 420.1	*-7.9%*	1 308.2	*+27.4%*	1 666.2
cows and heifers in milk	1 191.4	*-9.8%*	1 075.1	*+26.0%*	1 354.6
cows in calf but not in milk	228.7	*+1.9%*	233.2	*+33.6%*	311.7
Total heifers in calf (first calf) (b)	862.6	*+1.9%*	879.1	*-16.7%*	732.5
Dairy herd - total	700.4	*+1.5%*	710.7	*-24.9%*	533.8
two years old and over	*na*	*na*	466.1	*-33.9%*	308.2
under two years old	*na*	*na*	244.6	*-7.8%*	225.5
Beef herd - total	162.3	*+3.8%*	168.4	*+18.0%*	198.7
two years old and over	*na*	*na*	108.9	*+12.9%*	123.0
under two years old	*na*	*na*	59.5	*+27.3%*	75.7
Total bulls for service	84.5	*-9.8%*	76.2	*+6.4%*	81.1
Two years old and over	62.0	*-8.7%*	56.6	*+6.8%*	60.5
One year old and under two	22.5	*-12.9%*	19.6	*+5.3%*	20.6
Total other cattle and calves	7 580.2	*-5.9%*	7 131.8	*-7.2%*	6 616.5
Two years old and over - total	963.4	*-20.2%*	769.0	*-11.8%*	678.2
male (excluding bulls for service)	516.8	*-19.3%*	416.8	*-16.8%*	347.0
female intended for slaughter (c)	248.0	*-12.1%*	217.9	*-5.2%*	206.6
female for dairy or beef herd replacements	198.6	*-32.4%*	134.3	*-7.2%*	124.6
One year old and under two - total	3 041.3	*-7.3%*	2 818.5	*-8.4%*	2 581.0
male (excluding bulls for service)	1 235.3	*-5.1%*	1 172.9	*-11.1%*	1 042.6
female intended for slaughter (c)	866.5	*+3.0%*	892.8	*+0.1%*	893.4
female for dairy or beef herd replacements - total	939.5	*-19.9%*	752.8	*-14.3%*	645.0
female for dairy herd replacements (d)	*na*	*na*	508.0	*-22.6%*	393.1
female for beef herd replacements (d)	*na*	*na*	197.0	*+6.1%*	209.0
Six months old and under one year - total	1 876.1	*-0.0%*	1 876.0	*-9.7%*	1 694.9
male (including bull calves for service)	889.7	*+4.2%*	927.0	*-13.7%*	800.1
female	986.4	*-3.8%*	949.0	*-5.7%*	894.9
Under six months old - total	1 699.4	*-1.8%*	1 668.2	*-0.3%*	1 662.4
intended for slaughter as calves (e)	33.2	*-1.5%*	32.7	*-29.0%*	23.2
other: male (including bull calves for service)	813.0	*+1.9%*	828.6	*-2.3%*	809.2
other: female	853.3	*-5.4%*	806.9	*+2.9%*	830.0

Note

Please refer to notes at Table 4.6, pages 4-12 and 4-13.

Five- and ten-year comparisons
PIGS and SHEEP on Agricultural Holdings
United Kingdom (a)

Table 3.3

1981, 1986 and 1991 (at June Census)

thousands

	1981	% change 1986 1981	1986	% change 1991 1986	1991
TOTAL PIGS	**7 827.7**	**+1.4%**	**7 937.2**	**-4.3%**	**7 596.3**
Total breeding pigs	**966.8**	**-2.0%**	**947.4**	**-3.0%**	**918.9**
Breeding herd - total	836.4	-1.4%	824.3	-4.7%	785.7
mated sows and gilts - total	633.9	+1.3%	642.2	-3.6%	618.9
sows in pig	522.2	+2.2%	533.7	-4.1%	511.6
gilts in pig	111.6	-2.8%	108.5	-1.1%	107.3
other sows (being suckled or for breeding)	202.6	-10.1%	182.2	-8.4%	166.8
Boars for service	42.9	+3.1%	44.2	+0.9%	44.6
Gilts not yet in pig	87.4	-9.8%	78.8	+12.3%	88.5
Barren sows for fattening (b)	**10.7**	**+8.0%**	**11.6**	**-16.3%**	**9.7**
Total other pigs (c)	**6 850.3**	**+1.9%**	**6 978.2**	**-4.5%**	**6 667.6**
110kg and over (d)	90.1	-11.3%	79.9	-49.4%	40.5
80kg and under 110kg	638.5	-5.6%	602.6	+8.2%	652.2
50kg and under 80kg	1 776.4	+4.9%	1 862.7	-5.1%	1 766.9
20kg and under 50kg	2 226.5	+1.9%	2 269.9	-5.6%	2 142.8
under 20kg	2 118.7	+2.1%	2 163.1	-4.5%	2 065.2
TOTAL SHEEP AND LAMBS	**32 097.4**	**+15.3%**	**37 015.6**	**+17.8%**	**43 621.2**
Total sheep and lambs one year old and over	**16 469.2**	**+13.1%**	**18 631.5**	**+16.4%**	**21 679.4**
Breeding flock - total	15 270.6	+13.9%	17 397.9	+16.8%	20 325.6
ewes kept for breeding	12 527.7	+13.8%	14 251.6	+18.9%	16 944.4
two-tooth ewes (shearling ewes/gimmers)	2 742.9	+14.7%	3 146.3	+7.5%	3 381.3
Other sheep one year old and over - total	1 198.6	+2.9%	1 233.7	+9.7%	1 353.8
rams kept for service	357.7	+17.2%	419.2	+20.0%	503.0
draft and cast ewes, wethers and others	840.8	-3.1%	814.5	+4.5%	850.7
Lambs under one year old	**15 628.3**	**+17.6%**	**18 384.1**	**+19.4%**	**21 941.8**

Note

Please refer to notes at Table 4.11, page 4-22.

Five- and ten-year comparisons
POULTRY on Agricultural Holdings
United Kingdom (a)

Table 3.4   1981, 1986 and 1991(at June Census)   thousands

	1981	% change 1986 1981	1986	% change 1991 1986	1991
Total fowls	122 639	*-1.5%*	120 740	*+5.4%*	127 228
Growing pullets (day old to point of lay)	14 219	*-12.1%*	12 502	*-11.9%*	11 016
Total laying flock	44 473	*-14.3%*	38 096	*-12.7%*	33 273
in flock for: less than 12 months	31 737	*-6.2%*	29 778	*-11.8%*	26 260
12 months or more	12 736	*-34.7%*	8 318	*-15.7%*	7 013
Total breeding flock	6 117	*+3.5%*	6 334	*+14.3%*	7 238
Breeding hens (b)	4 901	*+5.8%*	5 187	*+10.2%*	5 715
Cocks and cockerels for breeding (b)	565	*+0.7%*	569	*-1.8%*	559
Table fowls	57 830	*+10.3%*	63 807	*+18.6%*	75 701
Total ducks and geese (c)	1 481	*+18.6%*	1 756	*+24.7%*	2 191
Ducks (d)	1 333	*+19.3%*	1 590	*+25.0%*	1 987
Geese (d)	148	*+1.6%*	150	*-2.4%*	146

Note

Please refer to notes at Table 4.16, page 4-32.

Table 3.5

Five- and ten-year comparisons
LABOUR FORCE on Agricultural Holdings
United Kingdom (a)
1981, 1986 and 1991 (at June Census)   thousands

	1981	% change 1986 / 1981	1986	% change 1991 / 1986	1991
TOTAL LABOUR FORCE	**709.9**	***-4.2%***	**679.9**	***-7.6%***	**627.9**
Total farmers,partners,directors (doing farm work) (b)	**293.6**	***-1.6%***	**288.9**	***-3.6%***	**278.6**
Whole-time - total	204.7	-3.5%	197.4	-10.0%	177.7
principal farmers and partners (c)	163.0	-4.9%	155.0	-9.6%	140.1
other partners and directors (d)	41.6	+1.8%	42.4	-11.3%	37.6
Part-time - total	89.0	+2.9%	91.5	+10.3%	100.9
principal farmers and partners (c)	67.6	+2.0%	68.9	+10.9%	76.5
other partners and directors (d)	21.4	+5.4%	22.6	+8.2%	24.4
Spouses of farmers,partners,directors	**74.6**	***+3.4%***	**77.1**	***-0.8%***	**76.5**
Salaried managers (d)	**7.9**	***+5.5%***	**8.3**	***-4.7%***	**7.9**
Total other workers (e)	**333.8**	***-8.5%***	**305.5**	***-13.3%***	**264.9**
Male	246.4	-9.3%	223.5	-15.2%	189.6
Female	87.4	-6.0%	82.1	-8.3%	75.3
Regular family workers - total	54.5	+5.8%	57.7	-17.0%	47.9
whole-time - total	35.0	+5.7%	37.0	-24.5%	27.9
male	29.8	+8.5%	32.3	-24.9%	24.2
female	5.3	-9.7%	4.7	-21.6%	3.7
part-time - total	19.5	+5.9%	20.7	-3.6%	19.9
male	12.5	+7.1%	13.4	-4.0%	12.9
female	7.0	+3.8%	7.3	-2.8%	7.1
Regular hired workers - total	182.2	-16.3%	152.6	-14.5%	130.4
whole-time - total	139.5	-19.8%	111.9	-18.2%	91.6
male	128.0	-20.5%	101.8	-21.0%	80.4
female	11.5	-12.0%	10.1	+10.7%	11.2
part-time - total	42.7	-4.8%	40.7	-4.6%	38.8
male	19.0	-1.2%	18.8	-2.9%	18.3
female	23.7	-7.7%	21.9	-6.2%	20.5
Seasonal or casual workers - total	97.0	-1.8%	95.3	-9.1%	86.6
male	57.1	+0.1%	57.2	-5.9%	53.8
female	39.9	-4.6%	38.1	-13.9%	32.8

Note

Please refer to notes at Table 4.21, pages 4-42 and 4-43.

Chapter 4
Annual reference tables; 1982 - 1991

Agricultural Land by TYPE OF USE
UNITED KINGDOM (a)
1982 - 1986 (at June Census)

Table 4.1

	1982	1983	1984	1985	1986
Total crops	**5 071.9**	**5 027.2**	**5 154.5**	**5 224.5**	**5 239.4**
+Bare fallow	55.5	97.1	41.6	40.8	48.0
=Total tillage	**5 127.4**	**5 124.3**	**5 196.1**	**5 265.3**	**5 287.4**
+Grasses under five years old	1 858.6	1 846.1	1 794.0	1 795.9	1 723.3
=Total arable land	**6 986.0**	**6 970.4**	**6 990.1**	**7 061.2**	**7 010.6**
+Grasses five years old and over	5 097.1	5 107.5	5 104.6	5 018.8	5 077.3
= Total tillage and grass	**12 083.1**	**12 077.8**	**12 094.7**	**12 080.0**	**12 087.9**
+Sole right rough grazing (b)	4 983.6	4 926.6	4 895.5	4 872.1	4 828.5
+Woodland on holdings	285.0	292.1	299.1	311.9	316.4
+All other land on holdings (c)	217.5	226.7	218.3	222.9	226.9
=TOTAL AREA ON HOLDINGS	**17 569.1**	**17 523.1**	**17 507.6**	**17 486.8**	**17 459.7**
Total crops	**5 071.9**	**5 027.2**	**5 154.5**	**5 224.5**	**5 239.4**
Total cereals (excluding maize)	**4 029.9**	**3 960.6**	**4 036.1**	**4 015.3**	**4 024.2**
Wheat	1 662.8	1 695.1	1 939.2	1 902.1	1 997.4
Barley - total	2 221.7	2 143.1	1 977.8	1 965.0	1 916.1
- winter	*na*	*na*	*na*	1 026.4	959.9
- spring	*na*	*na*	*na*	938.6	956.2
Oats	129.4	107.5	105.5	133.4	97.1
Mixed corn for threshing	9.6	8.4	7.6	7.3	6.7
Rye for threshing	6.4	6.5	6.0	7.5	6.9
Triticale	*na*	*na*	*na*	*na*	*na*
Total other arable crops not for stockfeeding	**582.2**	**629.3**	**677.4**	**708.0**	**704.0**
Potatoes (early and maincrop)	192.3	194.8	198.5	191.3	177.7
Sugar beet (not for stockfeeding)	203.6	199.3	199.1	205.4	205.0
Hops	5.8	5.7	5.2	4.9	4.4
Oilseed rape	174.5	222.3	268.6	295.6	299.1
Linseed	*na*	*na*	*na*	*na*	*na*
Other crops not for stockfeeding (d)	6.0	7.3	5.9	10.8	17.8
Total crops mainly for stockfeeding	**205.7**	**213.0**	**236.9**	**289.3**	**297.5**
Field beans	39.9	33.9	32.4	45.1	59.7
Peas for harvesting dry (e)	*na*	*na*	55.6	92.4	90.6
Other fodder crops (including maize for threshing)	165.8	150.4	148.9	151.8	147.2
Total horticultural crops	**254.1**	**224.3**	**204.2**	**211.9**	**213.6**
Vegetables for human consumption (f)	179.0	151.9	134.0	142.8	145.8
Orchards	*na*	*na*	39.3	38.7	37.9
Small fruit (g)	17.7	17.2	16.4	15.9	15.2
Hardy nursery stock, bulbs and flowers	12.5	12.2	12.3	12.3	12.4
Area under glass or plastic covered structures	2.2	2.2	2.2	2.2	2.3

Notes

(a) Estimates for minor holdings are included for England and Wales but not for Scotland and Northern Ireland. In 1991 the Scottish census was revised to exclude returns from about 2,500 holdings (net) which were reclassified as minor holdings. Retrospective revisions on this basis have been made from 1987 to 1990.

(b) 'Sole right rough grazing' before 1990 included rough grazing on land owned by the Northern Ireland Forestry Service.

(c) From 1989 includes 'Set-Aside Scheme' land in England and Wales.

Agricultural Land by TYPE OF USE
UNITED KINGDOM (a)
1987 - 1991 (at June Census)

thousand hectares

1987	1988	1989	1990	1991	
5 272.1	**5 255.0**	**5 137.4**	**5 013.4**	**4 956.0**	**Total crops**
41.6	58.0	65.3	64.0	63.9	+Bare fallow
5 313.8	**5 313.0**	**5 202.7**	**5 077.4**	**5 019.8**	**=Total tillage**
1 692.5	1 614.5	1 534.9	1 579.9	1 580.8	+Grasses under five years old
7 006.2	**6 927.5**	**6 737.6**	**6 657.4**	**6 600.6**	**=Total arable land**
5 110.0	5 159.0	5 248.9	5 263.3	5 267.0	+Grasses five years old and over
12 116.2	**12 086.5**	**11 986.5**	**11 920.7**	**11 867.6**	**=Total tillage and grass**
4 790.9	4 758.9	4 735.5	4 705.7	4 674.0	+Sole right rough grazing (b)
327.1	338.6	351.0	357.4	367.9	+Woodland on holdings
227.2	231.2	272.2	322.7	344.1	+All other land on holdings (c)
17 461.4	**17 415.2**	**17 345.2**	**17 306.5**	**17 253.6**	**=TOTAL AREA ON HOLDINGS**
5 272.1	**5 255.0**	**5 137.4**	**5 013.4**	**4 956.0**	**Total crops**
3 936.1	**3 897.1**	**3 873.2**	**3 657.4**	**3 499.8**	**Total cereals (excluding maize)**
1 993.9	1 885.7	2 082.6	2 013.1	1 980.5	Wheat
1 830.7	1 878.7	1 652.3	1 516.1	1 392.6	Barley - total
968.3	856.4	880.7	882.6	841.3	- winter
862.4	1 022.3	771.6	633.5	551.3	- spring
98.9	120.3	118.4	106.6	103.5	Oats
5.9	5.1	4.8	4.2	3.7	Mixed corn for threshing
6.7	7.4	7.4	8.2	8.6	Rye for threshing
na	*na*	7.7	9.2	11.0	Triticale
790.9	**756.0**	**720.3**	**806.0**	**916.4**	**Total other arable crops not for stockfeeding**
177.7	180.1	174.5	177.0	176.6	Potatoes (early and maincrop)
202.5	200.5	196.6	194.4	195.7	Sugar beet (not for stockfeeding)
4.2	4.0	3.9	3.9	3.7	Hops
387.6	347.2	320.7	389.9	439.9	Oilseed rape
na	*na*	17.4	33.7	91.9	Linseed
18.8	24.2	7.2	7.1	8.7	Other crops not for stockfeeding (d)
345.4	**392.7**	**336.1**	**342.0**	**336.3**	**Total crops mainly for stockfeeding**
91.0	153.7	129.3	139.2	130.9	Field beans
116.7	106.6	85.6	76.7	71.9	Peas for harvesting dry (e)
137.7	132.4	121.2	126.1	133.5	Other fodder crops (including maize for threshing)
199.8	**209.2**	**207.7**	**207.9**	**203.5**	**Total horticultural crops**
132.4	141.4	140.8	142.3	139.2	Vegetables for human consumption (f)
37.8	36.9	35.9	34.5	33.6	Orchards
14.9	15.2	15.0	15.2	14.6	Small fruit (g)
12.4	13.4	13.8	13.7	13.8	Hardy nursery stock, bulbs and flowers
2.2	2.3	2.3	2.3	2.3	Area under glass or plastic covered structures

(d) Before 1989 includes triticale and linseed.
(e) Includes peas harvested dry for both human consumption and stockfeeding.
(f) Excludes potatoes, peas for harvesting dry and mushrooms.
(g) Before 1989 excludes small fruit grown under orchard trees in England and Wales.
na not available.
Totals may not necessarily agree with the sum of their components due to rounding.

Agricultural Land by TYPE OF USE
ENGLAND (a)
1982 - 1986 (at June Census)

Table 4.2

	1982	1983	1984	1985	1986
Total crops	**4 271.4**	**4 237.6**	**4 338.7**	**4 395.6**	**4 416.4**
+Bare fallow	46.5	86.1	33.3	32.6	36.6
=Total tillage	**4 317.9**	**4 323.8**	**4 372.1**	**4 428.2**	**4 453.1**
+Grasses under five years old	960.3	956.4	917.8	921.6	887.9
=Total arable land	**5 278.2**	**5 280.1**	**5 289.9**	**5 349.8**	**5 340.9**
+Grasses five years old and over	3 170.6	3 151.0	3 143.2	3 064.3	3 066.7
=Total tillage and grass	**8 448.8**	**8 431.1**	**8 433.1**	**8 414.1**	**8 407.7**
+Sole right rough grazing	761.3	764.9	756.5	758.8	753.0
+Woodland on holdings	175.9	182.5	187.2	193.0	197.8
+All other land on holdings (b)	136.4	139.0	141.2	143.9	145.9
=TOTAL AREA ON HOLDINGS	**9 522.4**	**9 517.6**	**9 518.1**	**9 509.8**	**9 504.3**
Total crops	**4 271.4**	**4 237.6**	**4 338.7**	**4 395.6**	**4 416.4**
Total cereals (excluding maize)	**3 380.6**	**3 321.3**	**3 383.2**	**3 366.5**	**3 368.5**
Wheat	1 612.9	1 637.4	1 854.3	1 805.4	1 894.3
Barley - total	1 664.4	1 596.7	1 445.3	1 452.8	1 400.0
- winter	817.1	826.5	902.4	912.7	876.0
- spring	847.4	770.2	542.9	540.1	523.9
Oats	89.7	74.6	72.1	95.7	62.6
Mixed corn for threshing	7.3	6.3	5.7	5.2	4.9
Rye for threshing	6.4	6.4	5.8	7.4	6.7
Triticale	*na*	*na*	*na*	*na*	*na*
Total other arable crops not for stockfeeding	**525.6**	**570.0**	**609.8**	**629.7**	**631.7**
Potatoes (early and maincrop)	138.6	141.8	143.8	139.3	130.4
Sugar beet (not for stockfeeding)	203.5	199.2	199.0	205.2	204.9
Hops	5.8	5.7	5.2	4.9	4.3
Oilseed rape	172.6	218.0	256.9	271.4	276.1
Linseed	*na*	*na*	*na*	*na*	*na*
Other crops not for stockfeeding (c)	5.1	5.4	4.9	8.9	16.0
Total crops mainly for stockfeeding	**129.1**	**139.1**	**158.0**	**204.7**	**220.8**
Field beans	39.8	33.8	32.3	44.9	59.6
Peas for harvesting dry (d)	*na*	28.4	47.6	77.1	80.5
Other fodder crops (including maize for threshing)	89.3	76.9	78.1	82.7	80.7
Total horticultural crops	**236.1**	**207.3**	**187.8**	**194.7**	**195.4**
Vegetables for human consumption (e)	168.6	142.4	124.8	132.7	134.4
Orchards	40.3	38.4	37.0	36.4	35.8
Small fruit (f)	13.7	13.2	12.6	12.2	11.8
Hardy nursery stock, bulbs and flowers	11.4	11.2	11.2	11.2	11.4
Area under glass or plastic covered structures	2.1	2.1	2.1	2.1	2.1

Notes
(a) Includes estimates for minor holdings.
(b) From 1989 includes 'Set Aside Scheme' land.
(c) Before 1989 includes triticale and linseed.
(d) Includes peas for harvesting dry for both human consumption and stockfeeding.

Agricultural Land by TYPE OF USE
ENGLAND (a)
1987 - 1991 (at June Census)

thousand hectares

1987	1988	1989	1990	1991	
4 443.7	**4 435.4**	**4 361.8**	**4 265.7**	**4 222.1**	**Total crops**
31.5	47.1	43.3	34.5	32.4	+Bare fallow
4 475.2	**4 482.6**	**4 405.0**	**4 300.3**	**4 254.5**	**=Total tillage**
873.2	843.9	791.1	837.4	856.6	+Grasses under five years old
5 348.4	**5 326.5**	**5 196.2**	**5 137.7**	**5 111.1**	**=Total arable land**
3 067.6	3 073.4	3 111.3	3 106.9	3 088.1	+Grasses five years old and over
8 416.0	**8 399.9**	**8 307.5**	**8 244.5**	**8 199.2**	**=Total tillage and grass**
751.4	743.3	753.2	747.5	742.5	+Sole right rough grazing
201.6	207.2	213.6	219.1	224.4	+Woodland on holdings
146.9	147.8	183.9	228.3	254.4	+All other land on holdings (b)
9 515.9	**9 498.3**	**9 458.2**	**9 439.5**	**9 420.5**	**=TOTAL AREA ON HOLDINGS**
4 443.7	**4 435.4**	**4 361.8**	**4 265.7**	**4 222.1**	**Total crops**
3 295.9	**3 255.8**	**3 256.7**	**3 075.7**	**2 933.2**	**Total cereals (excluding maize)**
1 874.4	1 772.2	1 958.8	1 885.3	1 853.1	Wheat
1 347.3	1 395.4	1 204.7	1 101.8	990.3	Barley - total
865.3	764.7	792.7	794.0	754.8	- winter
482.0	630.7	412.0	307.9	235.4	- spring
63.4	77.4	76.7	69.8	68.8	Oats
4.2	3.7	3.3	3.0	2.7	Mixed corn for threshing
6.5	7.2	7.3	8.2	8.5	Rye for threshing
na	*na*	6.0	7.6	9.7	Triticale
694.6	**664.6**	**639.0**	**714.7**	**819.8**	**Total other arable crops not for stockfeeding**
131.0	134.7	132.2	134.6	135.2	Potatoes (early and maincrop)
202.4	200.4	196.4	194.2	195.5	Sugar beet (not for stockfeeding)
4.2	4.0	3.9	3.9	3.7	Hops
340.9	303.8	283.2	342.8	387.8	Oilseed rape
na	*na*	17.3	33.6	91.4	Linseed
16.1	21.6	6.0	5.6	6.2	Other crops not for stockfeeding (c)
270.6	**324.4**	**277.4**	**287.8**	**285.8**	**Total crops mainly for stockfeeding**
90.8	153.3	128.9	138.7	130.3	Field beans
103.0	97.3	80.0	71.9	68.1	Peas for harvesting dry (d)
76.8	73.8	68.5	77.1	87.4	Other fodder crops (including maize for threshing)
182.6	**190.7**	**188.7**	**187.5**	**183.2**	**Total horticultural crops**
121.7	129.7	128.9	128.8	125.6	Vegetables for human consumption (e)
35.7	34.9	33.9	32.5	31.7	Orchards
11.6	11.6	11.3	11.6	11.3	Small fruit (f)
11.5	12.3	12.4	12.4	12.5	Hardy nursery stock, bulbs and flowers
2.1	2.1	2.2	2.2	2.1	Area under glass or plastic covered structures

(e) Excludes potatoes, peas for harvesting dry and mushrooms.
(f) Before 1989 excludes small fruit grown under orchard trees.
na not available.
Totals may not necessarily agree with the sum of their components due to rounding.

Agricultural Land by TYPE OF USE
WALES (a)
1982 - 1986 (at June Census)

Table 4.3

	1982	1983	1984	1985	1986
Total crops	**92.9**	**86.9**	**87.4**	**88.1**	**85.4**
+Bare fallow	4.3	4.7	2.8	2.8	4.4
=Total tillage	**97.2**	**91.6**	**90.2**	**90.9**	**89.9**
+Grasses under five years old	166.2	167.2	164.5	173.8	168.7
=Total arable land	**263.4**	**258.9**	**254.7**	**264.7**	**258.5**
+Grasses five years old and over	845.0	861.6	866.0	854.1	871.0
=Total tillage and grass	**1 108.4**	**1 120.5**	**1 120.7**	**1 118.8**	**1 129.6**
+Sole right rough grazing	350.4	344.2	345.3	348.0	339.3
+Woodland on holdings	31.7	31.6	30.8	32.5	33.2
+All other land on holdings (b)	13.0	12.8	12.9	13.4	13.1
=TOTAL AREA ON HOLDINGS	**1 503.5**	**1 509.1**	**1 509.7**	**1 512.6**	**1 515.1**
Total crops	**92.9**	**86.9**	**87.4**	**88.1**	**85.4**
Total cereals (excluding maize)	**71.1**	**66.2**	**66.8**	**67.4**	**66.5**
Wheat	8.6	8.9	10.8	10.1	10.5
Barley - total	55.4	50.7	49.5	50.4	50.6
- winter	12.1	11.9	15.0	16.5	16.7
- spring	43.3	38.9	34.5	33.9	33.9
Oats	5.9	5.6	5.6	5.9	4.6
Mixed corn for threshing	1.1	0.9	0.8	0.8	0.6
Rye for threshing	0.1	0.1	0.2	0.1	0.2
Triticale	*na*	*na*	*na*	*na*	*na*
Total other arable crops not for stockfeeding	**6.6**	**6.8**	**6.7**	**6.8**	**6.5**
Potatoes (early and maincrop)	5.8	5.9	5.8	5.7	5.4
Sugar beet (not for stockfeeding)	0.1	0.1	0.2	0.2	0.2
Hops	..	..	..	..	..
Oilseed rape	0.2	0.3	0.4	0.5	0.6
Linseed	*na*	*na*	*na*	*na*	*na*
Other crops not for stockfeeding (c)	0.5	0.4	0.3	0.3	0.3
Total crops mainly for stockfeeding	**13.4**	**12.3**	**12.4**	**12.5**	**10.8**
Field beans	0.1	0.1	0.1	0.2	0.1
Peas for harvesting dry (d)	*na*	0.4	0.5	0.7	0.6
Other fodder crops (including maize for threshing)	13.3	11.8	11.8	11.7	10.1
Total horticultural crops	**1.8**	**1.6**	**1.5**	**1.5**	**1.6**
Vegetables for human consumption (e)	1.1	0.9	0.9	0.9	1.0
Orchards	0.1	0.1	0.1	0.1	0.1
Small fruit (f)	0.3	0.3	0.3	0.3	0.3
Hardy nursery stock, bulbs and flowers	0.2	0.2	0.2	0.2	0.2
Area under glass or plastic covered structures	..	..	..	..	..

Notes
(a) Includes estimates for minor holdings.
(b) From 1989 includes 'Set Aside Scheme' land.
(c) Before 1989 includes triticale and linseed.
(d) Includes peas for harvesting dry for both human consumption and stockfeeding.

Agricultural Land by TYPE OF USE
WALES (a)
1987 - 1991 (at June Census)

thousand hectares

1987	1988	1989	1990	1991	
85.4	**83.4**	**76.0**	**70.5**	**68.6**	**Total crops**
3.0	3.4	2.9	3.9	3.2	+Bare fallow
88.4	**86.8**	**78.9**	**74.4**	**71.9**	**=Total tillage**
169.0	161.3	146.8	148.6	149.3	+Grasses under five years old
257.4	**248.1**	**225.7**	**223.0**	**221.2**	**=Total arable land**
879.5	885.1	904.6	903.7	902.7	+Grasses five years old and over
1 136.9	**1 133.2**	**1 130.3**	**1 126.7**	**1 123.9**	**=Total tillage & grass**
334.2	337.3	332.6	335.0	335.5	+Sole right rough grazing
33.3	34.3	34.5	36.5	37.3	+Woodland on holdings
13.4	13.6	13.8	13.8	14.2	+All other land on holdings (b)
1 517.7	**1 518.4**	**1 511.1**	**1 512.0**	**1 511.0**	**=TOTAL AREA ON HOLDINGS**
85.4	**83.4**	**76.0**	**70.5**	**68.6**	**Total crops**
67.4	**66.7**	**61.6**	**56.0**	**53.1**	**Total cereals (excluding maize)**
10.7	9.8	11.2	11.3	11.8	Wheat
51.4	51.5	45.0	39.2	36.2	Barley - total
17.9	16.6	17.2	16.7	15.8	- winter
33.5	34.8	27.9	22.6	20.4	- spring
4.5	4.8	4.6	4.8	4.5	Oats
0.6	0.5	0.5	0.5	0.4	Mixed corn for threshing
0.1	0.1	0.1	0.1	0.0	Rye for threshing
na	*na*	0.2	0.2	0.2	Triticale
6.3	**5.9**	**5.3**	**5.5**	**6.1**	**Total other arable crops not for stockfeeding**
5.1	4.8	4.2	4.3	4.2	Potatoes (early and maincrop)
0.1	0.1	0.1	0.1	0.1	Sugar beet (not for stockfeeding)
..	..	..	*na*	*na*	Hops
0.7	0.7	0.6	0.7	1.1	Oilseed rape
na	*na*	0.1	0.1	0.5	Linseed
0.4	0.4	0.2	0.2	0.2	Other crops not for stockfeeding (c)
10.1	**9.1**	**7.6**	**7.3**	**7.9**	**Total crops mainly for stockfeeding**
0.2	0.4	0.5	0.5	0.6	Field beans
0.7	0.7	0.5	0.4	0.4	Peas for harvesting dry (d)
9.2	8.0	6.6	6.4	6.9	Other fodder crops (including maize for threshing)
1.6	**1.6**	**1.5**	**1.6**	**1.6**	**Total horticultural crops**
1.0	1.0	0.8	0.9	0.8	Vegetables for human consumption (e)
0.1	0.1	0.1	0.1	0.1	Orchards
0.3	0.3	0.3	0.2	0.3	Small fruit (f)
0.2	0.3	0.4	0.4	0.4	Hardy nursery stock, bulbs and flowers
..	..	..	..	..	Area under glass or plastic covered structures

(e) Excludes potatoes, peas for harvesting dry and mushrooms.
(f) Before 1989 excludes small fruit grown under orchard trees.
na not available.
.. Less than half the final digit shown.
Totals may not necessarily agree with the sum of their components due to rounding.

Agricultural Land by TYPE OF USE
SCOTLAND (a)
1982 - 1986 (at June Census)

Table 4.4

	1982	1983	1984	1985	1986
Total crops	**629.6**	**628.6**	**652.8**	**664.1**	**661.0**
+Bare fallow	4.7	6.2	5.4	5.5	6.9
=Total tillage	**634.2**	**634.8**	**658.2**	**669.5**	**667.9**
+Grasses under five years old	491.6	481.5	465.8	454.3	437.0
=Total arable land	**1 125.9**	**1 116.3**	**1 124.0**	**1 123.9**	**1 104.9**
+Grasses five years old and over	562.6	570.1	575.0	580.4	604.6
=Total tillage and grass	**1 688.5**	**1 686.4**	**1 699.0**	**1 704.3**	**1 709.5**
+Sole right rough grazing	3 682.8	3 627.1	3 605.2	3 580.0	3 552.7
+Woodland on holdings	65.7	66.2	69.8	75.0	74.7
+All other land on holdings	35.5	45.0	35.8	37.0	38.4
=TOTAL AREA ON HOLDINGS	**5 472.5**	**5 424.7**	**5 409.8**	**5 396.3**	**5 375.2**
Total crops	**629.6**	**628.6**	**652.8**	**664.1**	**661.0**
Total cereals (excluding maize)	**526.5**	**522.8**	**534.9**	**527.4**	**535.3**
Wheat	40.2	47.4	71.0	81.6	88.8
Barley - total	454.9	450.3	438.0	415.7	418.2
- winter	*na*	*na*	*na*	90.9	63.0
- spring	*na*	*na*	*na*	324.7	355.2
Oats	30.7	24.4	25.1	29.3	27.5
Mixed corn for threshing	0.6	0.7	0.8	0.8	0.8
Rye for threshing	*na*	*na*	*na*	*na*	*na*
Triticale	*na*	*na*	*na*	*na*	*na*
Total other arable crops not for stockfeeding	**36.1**	**39.6**	**46.5**	**58.0**	**53.6**
Potatoes (early and maincrop)	34.1	34.1	34.7	33.3	30.1
Sugar beet (not for stockfeeding)	*na*	*na*	*na*	*na*	*na*
Hops	*na*	*na*	*na*	*na*	*na*
Oilseed rape	1.6	4.0	11.0	23.2	22.1
Linseed	*na*	*na*	*na*	*na*	*na*
Other crops not for stockfeeding (b)	0.4	1.5	0.7	1.6	1.4
Total crops mainly for stockfeeding	**54.6**	**54.4**	**60.1**	**66.4**	**58.8**
Field beans	*na*	*na*	*na*	*na*	*na*
Peas for harvesting dry (c)	*na*	*na*	7.5	14.6	9.5
Other fodder crops	54.6	54.4	52.6	51.8	49.3
Total horticultural crops	**12.4**	**11.7**	**11.4**	**12.3**	**13.2**
Vegetables for human consumption (d)	8.0	7.4	7.2	8.1	9.2
Orchards	*na*	*na*	0.1	0.1	0.1
Small fruit	3.7	3.6	3.4	3.4	3.1
Hardy nursery stock, bulbs and flowers	0.7	0.7	0.7	0.7	0.7
Area under glass or plastic covered structures	0.1	0.1	0.1	0.1	0.1

Notes

(a) Excludes minor holdings. In 1991 the Scottish census was revised to exclude returns from about 2,500 holdings (net) which were reclassified as minor holdings. Retrospective revisions on this basis have been made from 1987 to 1990.

(b) Before 1987 includes triticale.

(c) Includes peas harvested dry for both human consumption and stockfeeding.

Agricultural Land by TYPE OF USE
SCOTLAND (a)
1987 - 1991 (at June Census)

thousand hectares

1987	1988	1989	1990	1991	
667.7	**662.5**	**632.8**	**612.2**	**599.2**	**Total crops**
7.2	7.4	19.2	25.6	28.3	+Bare fallow
674.8	**669.9**	**651.9**	**637.8**	**627.5**	**=Total tillage**
423.8	415.4	413.4	408.8	395.6	+Grasses under five years old
1 098.6	**1 085.3**	**1 065.3**	**1 046.7**	**1 023.1**	**=Total arable land**
623.6	633.5	651.8	670.0	688.3	+Grasses five years old and over
1 722.2	**1 718.8**	**1 717.0**	**1 716.6**	**1 711.4**	**=Total tillage and grass**
3 522.6	3 491.2	3 460.4	3 442.4	3 417.4	+Sole right rough grazing
81.0	85.2	90.6	89.1	93.3	+Woodland on holdings
38.2	40.0	44.3	52.2	47.5	+All other land on holdings
5 364.0	**5 335.3**	**5 312.4**	**5 300.4**	**5 269.6**	**=TOTAL AREA ON HOLDINGS**
667.7	**662.5**	**632.8**	**612.2**	**599.2**	**Total crops**
522.0	**525.6**	**507.0**	**480.4**	**467.5**	**Total cereals (excluding maize)**
103.7	98.7	108.0	111.0	109.7	Wheat
387.5	389.6	362.5	338.3	329.1	Barley - total
78.9	69.5	65.7	67.1	65.6	- winter
308.7	320.1	296.8	271.2	263.5	- spring
28.4	35.5	34.3	29.2	27.2	Oats
0.8	0.6	0.7	0.6	0.4	Mixed corn for threshing
na	*na*	*na*	*na*	*na*	Rye for threshing
1.6	1.1	1.5	1.4	1.1	Triticale
75.1	**71.3**	**64.8**	**73.7**	**78.6**	**Total other arable crops not for stockfeeding**
29.2	28.6	27.7	27.3	26.4	Potatoes (early and maincrop)
na	*na*	*na*	*na*	*na*	Sugar beet (not for stockfeeding)
na	*na*	*na*	*na*	*na*	Hops
45.0	41.6	36.1	45.2	49.9	Oilseed rape
na	*na*	*na*	*na*	*na*	Linseed
0.9	1.1	1.0	1.3	2.3	Other crops not for stockfeeding (b)
58.5	**52.4**	**46.9**	**42.7**	**37.8**	**Total crops mainly for stockfeeding**
na	*na*	*na*	*na*	*na*	Field beans
13.0	8.6	5.0	4.4	3.4	Peas for harvesting dry (c)
45.5	43.7	41.9	38.3	34.5	Other fodder crops
12.1	**13.3**	**14.0**	**15.4**	**15.3**	**Total horticultural crops**
8.3	9.2	9.7	11.2	11.5	Vegetables for human consumption (d)
0.1	0.1	..	..	0.1	Orchards
3.0	3.3	3.4	3.3	2.9	Small fruit
0.6	0.7	0.9	0.8	0.8	Hardy nursery stock, bulbs and flowers
..	0.1	..	..	..	Area under glass or plastic covered structures

(d) Excludes potatoes and also peas for harvesting dry.
na not available.
.. Less than half the final digit shown.
Totals may not necessarily agree with the sum of their components due to rounding.

Agricultural Land by TYPE OF USE
NORTHERN IRELAND (a)
1982 - 1986 (at June Census)

Table 4.5

	1982	1983	1984	1985	1986
Total crops	**78.0**	**74.1**	**75.6**	**76.6**	**76.5**
+Bare fallow	*na*	*na*	*na*	*na*	*na*
=**Total tillage**	**78.0**	**74.1**	**75.6**	**76.6**	**76.5**
+Grasses under five years old	240.5	241.0	245.9	246.2	229.8
=**Total arable land**	**318.5**	**315.1**	**321.5**	**322.8**	**306.3**
+Grasses five years old and over	518.8	524.8	520.4	520.0	534.8
=**Total tillage and grass**	**837.3**	**839.9**	**841.9**	**842.8**	**841.1**
+Sole right rough grazing (b)	189.1	190.3	188.4	185.4	183.5
+Woodland on holdings	11.7	11.8	11.4	11.4	10.7
+All other land on holdings	32.6	29.8	28.4	28.6	29.6
=**TOTAL AREA ON HOLDINGS**	**1 070.8**	**1 071.8**	**1 070.0**	**1 068.1**	**1 065.0**
Total crops	**78.0**	**74.1**	**75.6**	**76.6**	**76.5**
Total cereals (excluding maize)	**51.7**	**50.3**	**51.2**	**54.1**	**53.9**
Wheat	1.0	1.5	3.1	5.0	3.8
Barley - total	47.0	45.4	45.1	46.1	47.4
- winter	4.2	4.9	5.8	6.3	4.2
- spring	42.9	40.5	39.3	39.9	43.2
Oats	3.1	2.9	2.7	2.6	2.4
Mixed corn for threshing	0.6	0.5	0.4	0.4	0.3
Rye for threshing	*na*	*na*	*na*	*na*	*na*
Triticale	*na*	*na*	*na*	*na*	*na*
Total other arable crops not for stockfeeding	**13.9**	**12.9**	**14.4**	**13.5**	**12.1**
Potatoes (early and maincrop)	13.9	12.9	14.1	13.0	11.8
Sugar beet (not for stockfeeding)	*na*	*na*	*na*	*na*	*na*
Hops	*na*	*na*	*na*	*na*	*na*
Oilseed rape	*na*	*na*	0.3	0.5	0.3
Linseed	*na*	*na*	*na*	*na*	*na*
Other crops not for stockfeeding (c)	*na*	*na*	*na*	*na*	*na*
Total crops mainly for stockfeeding	**8.6**	**7.3**	**6.4**	**5.7**	**7.1**
Field beans	*na*	*na*	*na*	*na*	*na*
Peas for harvesting dry	*na*	*na*	*na*	*na*	*na*
Other fodder crops (c)	8.6	7.3	6.4	5.7	7.1
Total horticultural crops	**3.8**	**3.6**	**3.5**	**3.4**	**3.4**
Vegetables for human consumption (d)	1.3	1.2	1.2	1.1	1.3
Orchards	2.3	2.2	2.1	2.0	2.0
Small fruit	0.1	0.1	0.1	0.1	0.1
Hardy nursery stock, bulbs and flowers	0.2	0.1	0.1	0.2	0.1
Area under glass or plastic covered structures	..	..	..	..	..

Notes
(a) Excluding minor holdings.
(b) Before 1990 included rough grazing on land owned by the Northern Ireland Forest Service.
(c) Other crops not for stockfeeding are included with other fodder crops.
(d) Excludes potatoes and mushrooms.
na not available.

Agricultural Land by TYPE OF USE
NORTHERN IRELAND (a)
1987 - 1991 (at June Census)

thousand hectares

1987	1988	1989	1990	1991	
75.4	**73.7**	**66.9**	**64.9**	**66.0**	**Total crops**
na	*na*	*na*	*na*	*na*	+Bare fallow
75.4	**73.7**	**66.9**	**64.9**	**66.0**	**=Total tillage**
226.5	193.9	183.6	185.1	179.3	+Grasses under five years old
301.8	**267.6**	**250.4**	**250.0**	**245.3**	**=Total arable land**
539.4	567.0	581.2	582.7	587.9	+Grasses five years old and over
841.2	**834.6**	**831.6**	**832.8**	**833.2**	**=Total tillage and grass**
182.6	187.0	189.3	180.8	178.6	+Sole right rough grazing (b)
11.2	11.8	12.3	12.8	12.8	+Woodland on holdings
28.7	29.8	30.3	28.3	28.0	+All other land on holdings
1 063.7	**1 063.2**	**1 063.5**	**1 054.7**	**1 052.5**	**=TOTAL AREA ON HOLDINGS**
75.4	**73.7**	**66.9**	**64.9**	**66.0**	**Total crops**
52.4	**50.1**	**47.9**	**45.3**	**46.0**	**Total cereals (excluding maize)**
5.0	4.9	4.7	5.6	5.9	Wheat
44.4	42.3	40.1	36.7	37.0	Barley - total
6.2	5.6	5.2	4.9	5.0	- winter
38.2	36.7	35.0	31.9	32.0	- spring
2.6	2.6	2.8	2.9	2.9	Oats
0.3	0.3	0.3	0.1	0.2	Mixed corn for threshing
na	*na*	*na*	*na*	*na*	Rye for threshing
na	*na*	*na*	*na*	*na*	Triticale
13.3	**13.1**	**11.2**	**12.0**	**11.9**	**Total other arable crops not for stockfeeding**
12.4	12.0	10.3	10.8	10.8	Potatoes (early and maincrop)
na	*na*	*na*	*na*	*na*	Sugar beet (not for stockfeeding)
na	*na*	*na*	*na*	*na*	Hops
0.9	1.1	0.9	1.2	1.2	Oilseed rape
na	*na*	*na*	*na*	*na*	Linseed
na	*na*	*na*	*na*	*na*	Other crops not for stockfeeding (c)
6.1	**6.9**	**4.3**	**4.2**	**4.7**	**Total crops mainly for stockfeeding**
na	*na*	*na*	*na*	*na*	Field beans
na	*na*	*na*	*na*	*na*	Peas for harvesting dry
6.1	6.9	4.3	4.2	4.7	Other fodder crops (c)
3.6	**3.6**	**3.5**	**3.5**	**3.4**	**Total horticultural crops**
1.4	1.5	1.4	1.4	1.3	Vegetables for human consumption (d)
2.0	1.9	1.8	1.8	1.8	Orchards
0.1	0.1	0.1	0.1	0.1	Small fruit
0.1	0.1	0.1	0.1	0.1	Hardy nursery stock, bulbs and flowers
..	..	..	0.1	0.1	Area under glass or plastic covered structures

.. Less than half the final digit shown.
Totals may not necessarily agree with the sum of their components due to rounding.

CATTLE and CALVES on Agricultural Holdings
UNITED KINGDOM (a)
1982 - 1986 (at June Census)

Table 4.6

	1982	1983	1984	1985	1986
TOTAL CATTLE AND CALVES	**13 244.4**	**13 290.3**	**13 213.3**	**12 910.7**	**12 533.5**
Total breeding herd	**4 639.4**	**4 690.3**	**4 631.5**	**4 482.5**	**4 446.4**
Dairy herd - total	3 250.0	3 332.6	3 280.6	3 149.7	3 138.1
cows and heifers in milk	2 984.2	3 058.2	2 977.4	2 881.7	2 868.8
cows in calf but not in milk	265.8	274.4	303.2	268.1	269.3
Beef herd - total	1 389.4	1 357.7	1 350.9	1 332.8	1 308.2
cows and heifers in milk	1 161.5	1 132.3	1 123.7	1 103.2	1 075.1
cows in calf but not in milk	228.0	225.4	227.2	229.6	233.2
Total heifers in calf (first calf) (b)	**850.9**	**847.1**	**811.0**	**873.7**	**879.1**
Dairy herd - total	687.8	687.9	657.5	702.6	710.7
two years old and over	*na*	*na*	*na*	449.8	466.1
under two years old	*na*	*na*	*na*	252.8	244.6
Beef herd - total	163.2	159.1	153.5	171.1	168.4
two years old and over	*na*	*na*	*na*	108.1	108.9
under two years old	*na*	*na*	*na*	63.0	59.5
Total bulls for service	**83.9**	**83.3**	**80.3**	**78.0**	**76.2**
Two years old and over	60.9	60.7	59.0	57.3	56.6
One year old and under two	23.0	22.6	21.3	20.7	19.6
Total other cattle and calves	**7 670.1**	**7 669.6**	**7 690.4**	**7 476.5**	**7 131.8**
Two years old and over - total	937.2	903.9	905.5	851.5	769.0
male (excluding bulls for service)	510.5	489.3	488.1	462.2	416.8
female intended for slaughter (c)	242.0	239.5	246.2	226.4	217.9
female for dairy or beef herd replacements	184.7	175.1	171.2	162.9	134.3
One year old and under two - total	3 056.9	3 058.8	3 068.6	3 012.2	2 818.5
male (excluding bulls for service)	1 255.6	1 234.9	1 249.7	1 237.4	1 172.9
female intended for slaughter (c)	892.2	931.7	941.2	921.0	892.8
female for dairy or beef herd replacements - total	909.1	892.2	877.8	853.8	752.8
female for dairy herd replacements (d)	631.7	618.7	611.2	608.7	508.0
female for beef herd replacements (d)	217.0	214.1	209.1	190.1	197.0
Six months old and under one year - total	1 889.7	1 924.1	1 948.6	1 905.5	1 876.0
male (including bull calves for service)	888.6	911.1	927.6	933.0	927.0
female	1 001.1	1 013.0	1 021.0	972.5	949.0
Under six months old - total	1 786.4	1 782.7	1 767.7	1 707.3	1 668.2
intended for slaughter as calves (e)	33.1	38.6	39.3	33.8	32.7
other: male (including bull calves for service)	860.4	851.1	863.7	849.8	828.6
other: female	887.5	887.3	864.7	823.7	806.9

Notes

(a) Estimates for minor holdings are included for England and Wales but not for Scotland and Northern Ireland. In 1991 the Scottish census was revised to exclude returns from about 2,500 holdings (net) which were reclassified as minor holdings. Retrospective revisions on this basis have been made from 1987 to 1990.

(b) Following improved estimates there is a discontinuity in the series for heifers in calf in England between 1984 and 1985.

(c) In Scotland collected as "Not for breeding."

CATTLE and CALVES on Agricultural Holdings
UNITED KINGDOM (a)
1987 - 1991 (at June Census)

thousands

1987	1988	1989	1990	1991	
12 169.8	**11 883.8**	**11 975.4**	**12 059.4**	**11 866.0**	**TOTAL CATTLE AND CALVES**
4 387.4	**4 286.8**	**4 360.0**	**4 446.3**	**4 435.9**	**Total breeding herd**
3 042.4	2 911.8	2 864.7	2 847.2	2 769.7	Dairy herd - total
2 770.6	2 645.8	2 579.6	2 530.2	2 440.0	cows and heifers in milk
271.9	266.0	285.1	317.0	329.7	cows in calf but not in milk
1 345.0	1 375.0	1 495.2	1 599.1	1 666.2	Beef herd - total
1 092.2	1 120.5	1 220.3	1 299.4	1 354.6	cows and heifers in milk
252.8	254.5	274.9	299.7	311.7	cows in calf but not in milk
774.6	**834.3**	**793.4**	**756.6**	**732.5**	**Total heifers in calf (first calf) (b)**
598.5	612.2	566.2	529.3	533.8	Dairy herd - total
380.8	375.9	357.1	309.1	308.2	two years old and over
217.8	236.3	209.1	220.2	225.5	under two years old
176.1	222.1	227.2	227.3	198.7	Beef herd - total
111.8	134.0	143.7	139.8	123.0	two years old and over
64.3	88.1	83.6	87.5	75.7	under two years old
74.2	**75.0**	**78.4**	**82.2**	**81.1**	**Total bulls for service**
55.1	55.4	57.1	59.8	60.5	Two years old and over
19.1	19.6	21.2	22.4	20.6	One year old and under two
6 933.5	**6 687.7**	**6 743.7**	**6 774.3**	**6 616.5**	**Total other cattle and calves**
733.0	727.0	733.1	722.0	678.2	Two years old and over - total
390.0	393.2	396.8	391.2	347.0	male (excluding bulls for service)
213.0	202.6	206.1	204.4	206.6	female intended for slaughter (c)
130.0	131.2	130.2	126.3	124.6	female for dairy or beef herd replacements
2 752.2	2 672.3	2 599.4	2 654.3	2 581.0	One year old and under two - total
1 144.8	1 113.0	1 090.3	1 107.0	1 042.6	male (excluding bulls for service)
895.9	830.8	823.1	872.2	893.4	female intended for slaughter (c)
711.5	728.5	686.0	675.2	645.0	female for dairy or beef herd replacements - total
472.9	452.9	396.6	402.3	393.1	female for dairy herd replacements (d)
192.6	229.3	245.9	230.0	209.0	female for beef herd replacements (d)
1 845.7	1 704.0	1 774.4	1 746.4	1 694.9	Six months old and under one year - total
900.7	836.8	860.7	838.6	800.1	male (including bull calves for service)
945.0	867.2	913.7	907.8	894.9	female
1 602.7	1 584.4	1 636.7	1 651.6	1 662.4	Under six months old - total
30.5	26.7	20.8	22.1	23.2	intended for slaughter as calves (e)
795.9	791.3	808.9	811.4	809.2	other: male (including bull calves for service)
776.3	766.5	806.9	818.1	830.0	other: female

(d) Great Britain only. This category is not identified separately in Northern Ireland.

(e) England and Wales only. This category is not identified separately from other calves under six months old in Scotland and has not been separately identified in Northern Ireland from 1984.

na not available.

Totals may not necessarily agree with the sum of their components due to rounding.

CATTLE and CALVES on Agricultural Holdings
ENGLAND (a)
1982 - 1986 (at June Census)

Table 4.7

	1982	1983	1984	1985	1986
TOTAL CATTLE AND CALVES	**8 050.4**	**8 058.7**	**7 974.0**	**7 763.3**	**7 535.7**
Total breeding herd	**2 879.8**	**2 911.6**	**2 856.8**	**2 743.2**	**2 729.7**
Dairy herd - total	2 323.2	2 372.7	2 320.4	2 218.6	2 211.3
cows and heifers in milk	2 132.3	2 176.8	2 101.6	2 027.8	2 017.3
cows in calf but not in milk	190.9	195.9	218.8	190.9	194.0
Beef herd - total	556.7	538.9	536.4	524.5	518.4
cows and heifers in milk	451.5	435.5	429.4	420.4	412.6
cows in calf but not in milk	105.1	103.4	107.0	104.1	105.7
Total heifers in calf (first calf) (b)	**564.2**	**558.1**	**532.4**	**588.6**	**600.8**
Dairy herd - total	492.1	491.5	467.4	507.6	520.0
two years old and over	na	na	na	321.1	336.9
under two years old	na	na	na	186.5	183.0
Beef herd - total	72.1	66.6	65.0	81.0	80.9
two years old and over	na	na	na	48.3	51.5
under two years old	na	na	na	32.7	29.4
Total bulls for service	**44.2**	**43.6**	**41.9**	**40.2**	**39.5**
Two years old and over	31.9	31.7	30.8	29.5	29.2
One year old and under two	12.3	12.0	11.1	10.8	10.3
Total other cattle and calves	**4 562.1**	**4 545.4**	**4 542.9**	**4 391.3**	**4 165.6**
Two years old and over - total	555.5	530.6	524.3	490.0	433.8
male (excluding bulls for service)	272.3	258.5	254.9	239.0	214.5
female intended for slaughter	160.5	156.3	157.1	145.3	135.3
female for dairy or beef herd replacements	122.7	115.9	112.3	105.7	84.0
One year old and under two - total	1 829.1	1 814.6	1 804.2	1 762.3	1 619.8
male (excluding bulls for service)	689.2	675.8	676.3	664.1	624.5
female intended for slaughter	530.9	546.7	547.2	533.9	503.7
female for dairy or beef herd replacements - total	609.0	592.1	580.7	564.2	491.6
female for dairy herd replacements	482.4	470.3	459.5	459.1	379.7
female for beef herd replacements	126.6	121.8	121.1	105.1	111.9
Six months old and under one year - total	1 215.2	1 229.9	1 248.3	1 230.1	1 221.2
male (including bull calves for service)	565.9	577.9	587.8	598.1	602.2
female	649.3	652.0	660.5	632.0	619.1
Under six months old - total	962.3	970.3	966.2	908.9	890.8
intended for slaughter as calves	29.8	34.9	35.3	30.2	29.2
other: male (including bull calves for service)	463.3	462.4	471.4	453.5	444.4
other: female	469.2	473.0	459.4	425.2	417.2

Notes

(a) Includes estimates for minor holdings.

(b) Following improved estimates there is a discontinuity in the series for heifers in calf between 1984 and 1985.

na not available.

Totals may not necessarily agree with the sum of their components due to rounding.

CATTLE and CALVES on Agricultural Holdings
ENGLAND (a)
1987 - 1991 (at June Census)

thousands

1987	1988	1989	1990	1991	
7 302.5	7 052.3	7 091.4	7 097.4	6 881.6	**TOTAL CATTLE AND CALVES**
2 685.9	2 604.5	2 637.9	2 686.2	2 655.5	**Total breeding herd**
2 136.6	2 045.1	2 010.7	1 999.0	1 937.7	Dairy herd - total
1 939.9	1 853.4	1 801.8	1 766.7	1 695.2	cows and heifers in milk
196.6	191.7	208.9	232.3	242.4	cows in calf but not in milk
549.3	559.4	627.2	687.2	717.9	Beef herd - total
433.1	442.5	493.9	538.8	562.8	cows and heifers in milk
116.3	116.9	133.3	148.4	155.1	cows in calf but not in milk
512.5	552.2	528.1	500.9	487.7	**Total heifers in calf (first calf) (b)**
429.4	444.0	411.6	385.6	389.6	Dairy herd - total
267.8	266.1	254.6	217.9	218.5	two years old and over
161.6	177.9	157.1	167.7	171.1	under two years old
83.0	108.2	116.4	115.2	98.1	Beef herd - total
50.6	61.4	71.9	68.4	58.5	two years old and over
32.4	46.7	44.6	46.9	39.6	under two years old
38.4	38.4	40.3	42.4	41.2	**Total bulls for service**
28.0	28.1	29.1	30.5	30.3	Two years old and over
10.3	10.4	11.3	12.0	10.8	One year old and under two
4 065.8	3 857.2	3 885.1	3 867.9	3 697.1	**Total other cattle and calves**
410.9	405.4	404.9	392.0	366.5	Two years old and over - total
204.9	204.4	206.6	200.0	171.1	male (excluding bulls for service)
128.2	122.7	122.5	119.2	124.5	female intended for slaughter
77.8	78.2	75.8	72.8	70.8	female for dairy or beef herd replacements
1 591.7	1 540.3	1 481.1	1 510.6	1 433.1	One year old and under two - total
617.6	597.5	577.2	586.1	522.7	male (excluding bulls for service)
511.1	470.9	458.5	488.2	495.6	female intended for slaughter
463.0	471.8	445.4	436.2	414.8	female for dairy or beef herd replacements - total
351.1	337.7	296.7	301.7	295.2	female for dairy herd replacements
111.9	134.1	148.7	134.6	119.6	female for beef herd replacements
1 211.6	1 095.0	1 145.5	1 109.3	1 052.1	Six months old and under one year - total
587.2	533.8	551.3	525.2	483.0	male (including bull calves for service)
624.4	561.2	594.2	584.2	569.1	female
851.6	816.6	853.6	856.0	845.5	Under six months old - total
26.7	23.2	17.6	18.6	19.6	intended for slaughter as calves
424.8	411.3	421.7	419.0	407.0	other: male (including bull calves for service)
400.1	382.1	414.3	418.4	419.0	other: female

CATTLE and CALVES on Agricultural Holdings
WALES (a)
1982 - 1986 (at June Census)

Table 4.8

	1982	1983	1984	1985	1986
TOTAL CATTLE AND CALVES	**1 432.3**	**1 434.6**	**1 445.6**	**1 398.3**	**1 388.4**
Total breeding herd	**548.2**	**556.3**	**556.7**	**539.1**	**537.9**
Dairy herd - total	364.9	377.0	378.6	364.5	364.9
cows and heifers in milk	336.3	346.2	343.9	333.0	332.8
cows in calf but not in milk	28.6	30.8	34.7	31.5	32.1
Beef herd - total	183.3	179.4	178.2	174.6	172.9
cows and heifers in milk	153.0	148.6	146.5	142.8	139.4
cows in calf but not in milk	30.3	30.8	31.7	31.8	33.5
Total heifers in calf (first calf)	**83.6**	**83.1**	**81.9**	**84.0**	**96.2**
Dairy herd - total	64.1	64.2	63.4	65.5	74.0
two years old and over	na	na	na	44.8	50.7
under two years old	na	na	na	20.8	23.3
Beef herd - total	19.4	18.9	18.5	18.4	22.2
two years old and over	na	na	na	12.6	14.2
under two years old	na	na	na	5.8	8.1
Total bulls for service	**9.8**	**9.6**	**9.5**	**9.2**	**8.8**
Two years old and over	6.9	7.1	7.0	6.8	6.7
One year old and under two	3.0	2.6	2.5	2.4	2.2
Total other cattle and calves	**790.7**	**785.6**	**797.4**	**766.0**	**745.4**
Two years old and over - total	93.5	92.2	94.2	89.6	83.2
male (excluding bulls for service)	45.9	45.7	48.3	45.5	43.9
female intended for slaughter	23.1	22.9	23.9	22.1	22.1
female for dairy or beef herd replacements	24.4	23.6	22.0	22.0	17.3
One year old and under two - total	308.9	303.9	313.3	307.4	285.2
male (excluding bulls for service)	111.5	105.5	113.1	111.5	107.5
female intended for slaughter	87.1	92.5	91.5	91.4	86.3
female for dairy or beef herd replacements - total	110.2	105.9	108.7	104.5	91.5
female for dairy herd replacements	76.0	72.9	75.7	75.6	61.8
female for beef herd replacements	34.2	33.0	33.0	28.9	29.7
Six months old and under one year - total	187.4	189.4	193.3	182.6	188.5
male (including bull calves for service)	83.5	84.1	86.4	84.6	89.1
female	103.9	105.3	106.9	98.0	99.5
Under six months old - total	200.9	200.2	196.6	186.4	188.4
intended for slaughter as calves	3.3	3.7	4.0	3.6	3.5
other: male (including bull calves for service)	92.2	91.6	91.4	89.2	89.7
other: female	105.4	104.9	101.2	93.6	95.2

Notes

(a) Includes estimates for minor holdings.

na not available.

Totals may not necessarily agree with the sum of their components due to rounding.

CATTLE and CALVES on Agricultural Holdings
WALES (a)
1987 - 1991 (at June Census)

thousands

1987	1988	1989	1990	1991	
1 370.3	**1 341.7**	**1 354.5**	**1 362.9**	**1 343.4**	**TOTAL CATTLE AND CALVES**
531.6	**519.1**	**524.2**	**530.9**	**525.0**	**Total breeding herd**
352.8	336.8	329.9	327.0	317.3	Dairy herd - total
320.7	305.4	297.6	290.8	280.7	cows and heifers in milk
32.1	31.4	32.3	36.3	36.6	cows in calf but not in milk
178.8	182.3	194.3	203.9	207.7	Beef herd - total
143.4	146.6	156.5	163.2	166.1	cows and heifers in milk
35.4	35.6	37.8	40.7	41.6	cows in calf but not in milk
86.0	**91.9**	**86.8**	**79.6**	**78.3**	**Total heifers in calf (first calf)**
62.9	63.1	57.8	52.3	53.8	Dairy herd - total
42.4	41.7	38.8	33.2	34.0	two years old and over
20.5	21.4	19.0	19.2	19.8	under two years old
23.2	28.8	29.0	27.3	24.6	Beef herd - total
14.6	17.1	18.7	16.9	15.3	two years old and over
8.6	11.7	10.2	10.4	9.2	under two years old
8.7	**8.6**	**8.7**	**8.8**	**8.7**	**Total bulls for service**
6.7	6.7	6.6	6.6	6.7	Two years old and over
2.0	2.0	2.1	2.2	2.1	One year old and under two
744.0	**722.1**	**734.8**	**743.6**	**731.2**	**Total other cattle and calves**
82.4	84.4	85.1	84.4	85.3	Two years old and over - total
38.2	40.4	37.6	36.8	38.3	male (excluding bulls for service)
24.4	23.4	26.7	27.9	26.5	female intended for slaughter
19.9	20.6	20.8	19.7	20.5	female for dairy or beef herd replacements
291.6	282.8	278.6	295.8	285.0	One year old and under two - total
109.8	106.9	105.7	112.4	105.6	male (excluding bulls for service)
92.5	85.8	88.1	98.3	98.9	female intended for slaughter
89.3	90.1	84.8	85.1	80.5	female for dairy or beef herd replacements - total
58.1	55.3	47.8	50.0	49.3	female for dairy herd replacements
31.2	34.8	36.9	35.1	31.2	female for beef herd replacements
189.3	174.8	185.3	180.8	176.8	Six months old and under one year - total
89.2	83.6	85.4	83.8	82.1	male (including bull calves for service)
100.1	91.2	99.9	97.1	94.8	female
180.7	180.2	185.8	182.5	184.1	Under six months old - total
3.8	3.5	3.2	3.5	3.7	intended for slaughter as calves
86.4	85.5	88.8	86.5	86.3	other: male (including bull calves for service)
90.5	91.2	93.8	92.5	94.2	other: female

CATTLE and CALVES on Agricultural Holdings
SCOTLAND (a)
1982 - 1986 (at June Census)

Table 4.9

	1982	1983	1984	1985	1986
TOTAL CATTLE AND CALVES	**2 332.9**	**2 325.0**	**2 287.0**	**2 235.1**	**2 138.9**
Total breeding herd	**734.7**	**734.8**	**723.3**	**706.2**	**689.2**
Dairy herd - total	282.3	289.4	283.1	273.1	270.0
cows and heifers in milk	256.1	261.9	254.8	246.6	245.7
cows in calf but not in milk	26.2	27.5	28.3	26.4	24.3
Beef herd - total	452.4	445.4	440.1	433.1	419.2
cows and heifers in milk	388.8	380.5	377.3	366.0	352.6
cows in calf but not in milk	63.6	64.9	62.8	67.1	66.6
Total heifers in calf (first calf)	**122.3**	**123.0**	**116.7**	**117.8**	**109.8**
Dairy herd - total	74.4	73.9	72.0	72.3	68.2
two years old and over	na	na	na	55.2	53.4
under two years old	na	na	na	17.1	14.7
Beef herd - total	47.8	49.1	44.7	45.5	41.6
two years old and over	na	na	na	32.1	29.3
under two years old	na	na	na	13.4	12.3
Total bulls for service	**20.5**	**20.4**	**19.1**	**18.6**	**17.8**
Two years old and over	16.0	15.8	15.1	14.6	14.0
One year old and under two	4.5	4.6	4.1	3.9	3.8
Total other cattle and calves	**1 455.4**	**1 446.8**	**1 428.0**	**1 392.6**	**1 322.1**
Two years old and over - total	122.8	110.3	110.4	99.1	92.9
male (excluding bulls for service)	72.6	63.6	61.2	56.3	49.8
female intended for slaughter (b)	25.7	24.1	25.0	20.0	21.7
female for dairy or beef herd replacements	24.5	22.6	24.3	22.9	21.5
One year old and under - two total	598.9	609.7	602.9	585.1	558.5
male (excluding bulls for service)	287.4	281.2	280.1	277.0	259.2
female intended for slaughter (b)	181.9	193.8	191.9	178.0	177.3
female for dairy or beef herd replacements - total	129.6	134.7	131.0	130.1	122.0
female for dairy herd replacements	73.3	75.4	75.9	74.1	66.5
female for beef herd replacements	56.3	59.3	55.0	56.1	55.5
Six months old and under one year - total	344.6	343.0	340.1	330.0	311.8
male (including bull calves for service)	168.6	166.7	167.8	165.1	156.4
female	176.0	176.3	172.3	165.0	155.5
Under six months old - total	389.1	383.8	374.5	378.3	358.9
intended for slaughter as calves (c)	na	na	na	na	na
other: male (including bull calves for service)	190.8	187.6	186.1	189.3	179.7
other: female	198.3	196.2	188.4	189.0	179.2

Notes
(a) Excludes estimates for minor holdings. In 1991 the Scottish census was revised to exclude returns from about 2,500 holdings (net) which were reclassified as minor holdings. Retrospective revisions on this basis have been made from 1987 to 1990.
(b) Collected as "Not for breeding."
(c) This category is not identified separately from other calves under six months old.
na not available.
Totals may not necessarily agree with the sum of their components due to rounding.

CATTLE and CALVES on Agricultural Holdings
SCOTLAND (a)
1987 - 1991 (at June Census)

thousands

1987	1988	1989	1990	1991	
2 067.6	**2 050.6**	**2 062.3**	**2 092.7**	**2 107.9**	**TOTAL CATTLE AND CALVES**
685.4	**679.4**	**696.2**	**712.9**	**727.0**	**Total breeding herd**
264.2	251.0	246.8	243.6	240.7	Dairy herd - total
240.3	226.5	221.2	215.5	210.9	cows and heifers in milk
23.9	24.5	25.6	28.1	29.7	cows in calf but not in milk
421.2	428.4	449.4	469.3	486.3	Beef herd - total
352.1	358.5	377.2	392.6	407.9	cows and heifers in milk
69.0	69.8	72.2	76.7	78.4	cows in calf but not in milk
106.5	**115.1**	**109.1**	**106.3**	**99.4**	**Total heifers in calf (first calf)**
61.9	60.5	56.5	52.1	50.2	Dairy herd - total
47.8	45.9	43.7	39.6	37.3	two years old and over
14.1	14.6	12.8	12.5	12.9	under two years old
44.6	54.6	52.6	54.1	49.2	Beef herd - total
31.5	37.1	36.1	37.3	33.6	two years old and over
13.1	17.5	16.5	16.9	15.6	under two years old
17.4	**17.4**	**18.0**	**18.7**	**18.5**	**Total bulls for service**
13.7	13.8	13.9	14.5	14.6	Two years old and over
3.7	3.7	4.0	4.1	3.9	One year old and under two
1 258.3	**1 238.8**	**1 239.1**	**1 254.9**	**1 263.0**	**Total other cattle and calves**
89.2	81.3	82.2	84.2	80.6	Two years old and over - total
47.3	43.3	41.9	42.3	39.5	male (excluding bulls for service)
20.8	16.1	16.9	18.6	18.0	female intended for slaughter (b)
21.1	21.8	23.4	23.3	23.1	female for dairy or beef herd replacements
524.6	511.7	500.5	506.6	509.0	One year old and under two - total
243.2	235.6	233.5	230.8	228.3	male (excluding bulls for service)
168.1	155.8	154.8	164.7	174.0	female intended for slaughter (b)
113.2	120.3	112.2	111.0	106.7	female for dairy or beef herd replacements - total
63.7	60.0	52.0	50.7	48.5	female for dairy herd replacements
49.5	60.3	60.2	60.4	58.2	female for beef herd replacements
294.9	287.2	289.1	287.7	286.4	Six months old and under one year- total
148.1	144.0	143.6	142.7	141.6	male (including bull calves for service)
146.8	143.2	145.5	145.0	144.7	female
349.6	358.7	367.3	376.4	387.0	Under six months old - total
na	*na*	*na*	*na*	*na*	intended for slaughter as calves (c)
174.3	178.9	182.4	187.1	192.0	other: male (including bull calves for service)
175.3	179.8	184.9	189.3	195.0	other: female

CATTLE and CALVES on Agricultural Holdings
NORTHERN IRELAND (a)
1982 - 1986 (at June Census)

Table 4.10

	1982	1983	1984	1985	1986
TOTAL CATTLE AND CALVES	**1 428.9**	**1 471.9**	**1 506.6**	**1 514.0**	**1 470.6**
Total breeding herd	**476.7**	**487.6**	**494.7**	**494.1**	**489.6**
Dairy herd - total	279.6	293.5	298.5	293.5	291.9
cows and heifers in milk	259.5	273.3	277.1	274.2	273.0
cows in calf but not in milk	20.2	20.2	21.4	19.3	18.9
Beef herd - total	197.0	194.1	196.2	200.6	197.7
cows and heifers in milk	168.1	167.7	170.5	174.0	170.4
cows in calf but not in milk	29.0	26.3	25.7	26.6	27.3
Total heifers in calf (first calf)	**80.9**	**82.8**	**80.1**	**83.3**	**72.3**
Dairy herd - total	57.1	58.3	54.7	57.1	48.5
two years old and over	na	na	na	28.7	25.0
under two years old	na	na	na	28.4	23.5
Beef herd - total	23.8	24.5	25.4	26.1	23.8
two years old and over	na	na	na	15.0	14.0
under two years old	na	na	na	11.1	9.8
Total bulls for service	**9.4**	**9.7**	**9.8**	**10.0**	**10.0**
Two years old and over	6.2	6.2	6.2	6.4	6.7
One year old and under two	3.2	3.5	3.6	3.6	3.3
Total other cattle and calves	**862.0**	**891.8**	**922.1**	**926.6**	**898.6**
Two years old and over - total	165.3	170.9	176.5	172.8	159.0
male (excluding bulls for service)	119.6	121.5	123.7	121.3	108.7
female intended for slaughter	32.7	36.3	40.2	39.1	38.8
female for dairy or beef herd replacements	13.0	13.1	12.6	12.4	11.5
One year old and under two - total	320.1	330.6	348.2	357.5	355.1
male (excluding bulls for service)	167.5	172.5	180.2	184.8	181.7
female intended for slaughter	92.3	98.7	110.6	117.7	125.6
female for dairy or beef herd replacements - total	60.4	59.4	57.5	55.0	47.8
female for dairy herd replacements	na	na	na	na	na
female for beef herd replacements	na	na	na	na	na
Six months old and under one year - total	142.4	161.9	166.9	162.7	154.4
male (including bull calves for service)	70.6	82.4	85.6	85.1	79.5
female	71.8	79.4	81.3	77.6	75.0
Under six months old - total	234.1	228.5	230.4	233.7	230.1
intended for slaughter as calves (b)	5.4	5.7	na	na	na
other: male (including bull calves for service)	114.0	109.5	114.8	117.8	114.8
other: female	114.7	113.2	115.7	115.9	115.3

Notes

(a) Excludes estimates for minor holdings.
(b) From 1984 this category is not identified separately from other calves under six months old.
na not available.
Totals may not necessarily agree with the sum of their components due to rounding.

CATTLE and CALVES on Agricultural Holdings
NORTHERN IRELAND (a)
1987 - 1991 (at June Census)

thousands

1987	1988	1989	1990	1991	
1 429.4	**1 439.2**	**1 467.2**	**1 506.4**	**1 533.2**	**TOTAL CATTLE AND CALVES**
484.5	**483.9**	**501.6**	**516.3**	**528.3**	**Total breeding herd**
288.8	278.9	277.3	277.6	274.1	Dairy herd - total
269.6	260.5	259.0	257.2	253.2	cows and heifers in milk
19.2	18.5	18.2	20.4	20.9	cows in calf but not in milk
195.8	205.0	224.4	238.6	254.3	Beef herd - total
163.7	172.8	192.7	204.8	217.7	cows and heifers in milk
32.1	32.1	31.7	33.9	36.6	cows in calf but not in milk
69.6	**75.1**	**69.5**	**69.8**	**67.1**	**Total heifers in calf (first calf)**
44.3	44.6	40.3	39.2	40.2	Dairy herd - total
22.8	22.2	20.0	18.4	18.5	two years old and over
21.5	22.4	20.3	20.8	21.7	under two years old
25.3	30.5	29.2	30.6	26.9	Beef herd - total
15.1	18.3	16.9	17.3	15.5	two years old and over
10.2	12.2	12.3	13.3	11.4	under two years old
9.8	**10.5**	**11.4**	**12.4**	**12.7**	**Total bulls for service**
6.7	6.9	7.6	8.2	8.9	Two years old and over
3.0	3.6	3.8	4.2	3.8	One year old and under two
865.5	**869.7**	**884.6**	**907.9**	**925.1**	**Total other cattle and calves**
150.5	156.0	160.9	161.3	145.8	Two years old and over - total
99.6	105.1	110.6	112.1	98.1	male (excluding bulls for service)
39.7	40.4	40.1	38.7	37.6	female intended for slaughter
11.2	10.6	10.2	10.5	10.1	female for dairy or beef herd replacements
344.3	337.6	339.2	341.4	353.8	One year old and under two - total
174.1	173.0	173.9	177.6	186.0	male (excluding bulls for service)
124.2	118.4	121.7	121.0	124.8	female intended for slaughter
46.0	46.3	43.6	42.8	43.0	female for dairy or beef herd replacements - total
na	na	na	na	na	female for dairy herd replacements
na	na	na	na	na	female for beef herd replacements
149.9	147.1	154.5	168.6	179.7	Six months old and under one year - total
76.2	75.5	80.4	86.9	93.4	male (including bull calves for service)
73.7	71.6	74.1	81.6	86.3	female
220.8	229.0	229.9	236.7	245.8	Under six months old - total
na	na	na	na	na	intended for slaughter as calves (b)
110.3	115.6	116.0	118.8	124.0	other: male (including bull calves for service)
110.4	113.4	113.9	117.9	121.8	other: female

PIGS and SHEEP on Agricultural Holdings
UNITED KINGDOM (a)
1982 - 1986 (at June Census)

Table 4.11

	1982	1983	1984	1985	1986
TOTAL PIGS	**8 023.1**	**8 173.9**	**7 689.2**	**7 865.0**	**7 937.2**
Total breeding pigs	**997.5**	**983.0**	**919.4**	**951.7**	**947.4**
Breeding herd - total	863.9	855.9	800.1	828.4	824.3
mated sows and gilts - total	664.3	652.1	622.3	641.9	642.2
sows in pig	542.7	542.1	517.8	530.3	533.7
gilts in pig	121.6	110.0	104.5	111.6	108.5
other sows (being suckled or for breeding)	199.7	203.8	177.8	186.5	182.2
Boars for service	44.7	45.0	42.1	43.7	44.2
Gilts not yet in pig	88.8	82.1	77.2	79.6	78.8
Barren sows for fattening (b)	**11.8**	**14.9**	**12.4**	**11.6**	**11.6**
Total other pigs (c)	**7 013.9**	**7 176.0**	**6 757.5**	**6 901.7**	**6 978.2**
110kg and over (d)	116.6	100.4	91.3	89.3	79.9
80kg and under 110kg	629.5	604.5	598.5	588.6	602.6
50kg and under 80kg	1 823.6	1 867.9	1 787.3	1 813.0	1 862.7
20kg and under 50kg	2 281.4	2 362.1	2 197.9	2 260.4	2 269.9
under 20kg	2 162.8	2 241.1	2 082.4	2 150.6	2 163.1
TOTAL SHEEP AND LAMBS	**33 067.2**	**34 069.4**	**34 802.0**	**35 627.7**	**37 015.6**
Total sheep and lambs one year old and over	**17 022.9**	**17 457.1**	**17 722.1**	**18 061.9**	**18 631.5**
Breeding flock - total	15 779.6	16 243.3	16 540.2	16 877.7	17 397.9
ewes kept for breeding	12 909.0	13 310.4	13 648.4	13 893.4	14 251.6
two-tooth ewes (shearling ewes/gimmers)	2 870.6	2 932.9	2 891.8	2 984.3	3 146.3
Other sheep one year old and over - total	1 243.3	1 213.9	1 181.9	1 184.2	1 233.7
rams kept for service	365.9	382.5	393.5	405.9	419.2
draft and cast ewes, wethers and others	877.4	831.3	788.4	778.3	814.5
Lambs under one year old	**16 044.3**	**16 612.2**	**17 079.9**	**17 565.7**	**18 384.1**

Notes

(a) Estimates for minor holdings are included for England and Wales but not for Scotland and Northern Ireland. In 1991 the Scottish census was revised to exclude returns from about 2,500 holdings (net) which were reclassified as minor holdings. Retrospective revisions on this basis have been made from 1987 to 1990.

(b) Great Britain only. Barren sows for fattening are included with "All other pigs weighing 110 kg and over" in Northern Ireland.

(c) Weights are liveweight.

(d) Includes barren sows for fattening in Northern Ireland.

na not available.

Totals may not necessarily agree with the sum of their components due to rounding.

PIGS and SHEEP on Agricultural Holdings
UNITED KINGDOM (a)
1987 - 1991 (at June Census)

thousands

1987	1988	1989	1990	1991	
7 943.2	**7 981.6**	**7 508.7**	**7 449.1**	**7 596.3**	**TOTAL PIGS**
945.3	**920.9**	**872.3**	**897.3**	**918.9**	**Total breeding pigs**
820.5	804.6	757.0	768.7	785.7	Breeding herd - total
640.8	625.6	591.2	603.2	618.9	mated sows and gilts - total
533.6	524.4	494.5	494.6	511.6	sows in pig
107.2	101.3	96.6	108.6	107.3	gilts in pig
179.6	179.0	165.9	165.4	166.8	other sows (being suckled or for breeding)
44.2	43.1	41.6	43.3	44.6	Boars for service
80.7	73.1	73.7	85.3	88.5	Gilts not yet in pig
11.2	**11.7**	**9.7**	**9.8**	**9.7**	**Barren sows for fattening (b)**
6 986.7	**7 049.1**	**6 626.7**	**6 542.0**	**6 667.6**	**Total other pigs (c)**
66.8	62.2	45.4	50.8	40.5	110kg and over (d)
614.9	621.4	612.9	620.0	652.2	80kg and under 110kg
1 890.6	1 898.6	1 778.9	1 740.7	1 766.9	50kg and under 80kg
2 253.8	2 263.0	2 131.4	2 108.8	2 142.8	20kg and under 50kg
2 160.5	2 203.9	2 058.1	2 021.7	2 065.2	under 20kg
38 755.8	**41 007.0**	**42 987.9**	**43 798.7**	**43 621.2**	**TOTAL SHEEP AND LAMBS**
19 374.4	**20 411.1**	**21 423.5**	**21 775.8**	**21 679.4**	**Total sheep and lambs one year old and over**
18 122.6	19 076.7	20 039.1	20 410.7	20 325.6	Breeding flock - total
14 835.7	15 520.7	16 205.1	16 760.2	16 944.4	ewes kept for breeding
3 286.9	3 556.0	3 834.0	3 650.5	3 381.3	two-tooth ewes (shearling ewes/gimmers)
1 251.8	1 334.4	1 384.3	1 365.1	1 353.8	Other sheep one year old and over - total
436.7	460.8	490.3	500.3	503.0	rams kept for service
815.1	873.6	894.0	864.8	850.7	draft and cast ewes, wethers and others
19 381.4	**20 595.9**	**21 564.4**	**22 022.9**	**21 941.8**	**Lambs under one year old**

PIGS and SHEEP on Agricultural Holdings
ENGLAND (a)
1982 - 1986 (at June Census)

Table 4.12

	1982	1983	1984	1985	1986
TOTAL PIGS	**6 782.5**	**6 933.4**	**6 540.2**	**6 703.9**	**6 776.2**
Total breeding pigs	**840.9**	**832.8**	**783.7**	**810.4**	**808.6**
Breeding herd - total	726.6	722.7	680.4	704.3	703.6
mated sows and gilts - total	561.2	554.0	532.5	547.8	551.4
sows in pig	460.1	460.3	443.3	452.7	458.4
gilts in pig	101.1	93.7	89.3	95.2	92.9
other sows (being suckled or for breeding)	165.4	168.6	147.8	156.4	152.2
Boars for service	37.7	38.1	35.8	37.1	37.8
Gilts not yet in pig	76.6	72.1	67.5	69.1	67.1
Barren sows for fattening	**10.8**	**13.0**	**11.1**	**10.2**	**10.1**
Total other pigs (b)	**5 930.8**	**6 087.6**	**5 745.5**	**5 883.2**	**5 957.5**
110kg and over	102.8	84.2	75.5	74.6	66.8
80kg and under 110kg	526.6	506.0	501.3	496.8	513.7
50kg and under 80kg	1 523.5	1 577.9	1 509.3	1 534.2	1 575.0
20kg and under 50kg	1 927.7	2 003.1	1 866.0	1 932.6	1 941.1
under 20kg	1 850.2	1 916.3	1 793.3	1 845.1	1 860.8
TOTAL SHEEP AND LAMBS	**15 456.4**	**16 000.6**	**16 206.3**	**16 549.0**	**17 338.6**
Total sheep and lambs one year old and over	**7 536.7**	**7 782.4**	**7 851.9**	**8 006.6**	**8 343.2**
Breeding flock - total	6 977.2	7 229.1	7 323.0	7 467.7	7 773.9
ewes kept for breeding	5 764.3	5 963.6	6 101.9	6 197.9	6 387.0
two-tooth ewes (shearling ewes/gimmers)	1 213.0	1 265.5	1 221.2	1 269.8	1 386.9
Other sheep one year old and over - total	559.5	553.3	528.8	538.9	569.3
rams kept for service	159.5	168.6	172.9	179.9	186.8
draft and cast ewes, wethers and others total	400.0	384.7	355.9	359.0	382.5
draft and cast ewes	209.1	213.1	199.9	191.4	201.3
wethers and other sheep	190.9	171.5	156.0	167.6	181.2
Lambs under one year old	**7 919.7**	**8 218.3**	**8 354.4**	**8 542.4**	**8 995.4**

Notes

(a) Includes estimates for minor holdings.

(b) Weights are liveweight.

Totals may not necessarily agree with the sum of their components due to rounding.

PIGS and SHEEP on Agricultural Holdings
ENGLAND (a)
1987- 1991 (at June Census)

thousands

1987	1988	1989	1990	1991	
6 779.4	**6 770.1**	**6 345.6**	**6 308.3**	**6 411.8**	**TOTAL PIGS**
806.9	**778.0**	**738.1**	**756.5**	**773.8**	**Total breeding pigs**
700.0	680.1	639.0	646.9	663.3	Breeding herd - total
549.2	530.6	500.5	509.6	523.7	mated sows and gilts - total
457.5	445.2	418.6	417.3	434.2	sows in pig
91.7	85.5	81.9	92.3	89.5	gilts in pig
150.7	149.4	138.5	137.3	139.6	other sows (being suckled or for breeding)
37.8	36.5	35.2	36.5	37.9	Boars for service
69.2	61.5	63.9	73.1	72.6	Gilts not yet in pig
9.9	**10.5**	**8.9**	**8.7**	**8.9**	**Barren sows for fattening**
5 962.6	**5 981.6**	**5 598.6**	**5 543.1**	**5 629.1**	Total other pigs (b)
56.1	50.0	37.6	43.8	34.1	110kg and over
519.1	518.9	503.3	512.9	544.8	80kg and under 110kg
1 603.4	1 597.3	1 497.1	1 473.2	1 475.6	50kg and under 80kg
1 926.4	1 926.4	1 801.9	1 785.1	1 809.3	20kg and under 50kg
1 857.5	1 889.0	1 758.8	1 728.2	1 765.2	under 20kg
18 244.4	**19 387.8**	**20 540.7**	**20 775.9**	**20 438.7**	**TOTAL SHEEP AND LAMBS**
8 739.3	**9 301.8**	**9 896.6**	**10 016.6**	**9 834.6**	**Total sheep and lambs one year old and over**
8 151.4	8 672.7	9 241.3	9 375.0	9 206.5	Breeding flock - total
6 681.5	7 056.4	7 457.1	7 745.6	7 782.7	ewes kept for breeding
1 469.9	1 616.3	1 784.2	1 629.4	1 423.8	two-tooth ewes (shearling ewes/gimmers)
587.9	629.1	655.3	641.6	628.2	Other sheep one year old and over - total
198.5	210.4	228.0	233.7	232.1	rams kept for service
389.4	418.8	427.2	407.9	396.1	draft and cast ewes, wethers and others total
213.3	217.0	225.2	221.9	209.9	draft and cast ewes
176.1	201.8	202.0	186.0	186.2	wethers and other sheep
9 505.1	**10 086.0**	**10 644.1**	**10 759.3**	**10 604.1**	**Lambs under one year old**

PIGS and SHEEP on Agricultural Holdings
WALES (a)
1982 - 1986 (at June Census)

Table 4.13

	1982	1983	1984	1985	1986
TOTAL PIGS	**139.3**	**147.3**	**128.4**	**125.8**	**129.9**
Total breeding pigs	**20.8**	**20.2**	**17.7**	**19.1**	**18.8**
Breeding herd - total	17.5	17.4	14.9	16.1	16.2
mated sows and gilts - total	13.6	13.1	11.1	12.4	12.3
sows in pig	10.5	10.8	9.0	10.1	10.0
gilts in pig	3.1	2.4	2.1	2.3	2.3
other sows (being suckled or for breeding)	3.9	4.3	3.9	3.7	3.9
Boars for service	1.1	1.1	0.9	1.0	1.0
Gilts not yet in pig	2.2	1.8	1.8	2.0	1.7
Barren sows for fattening	**0.3**	**0.7**	**0.7**	**0.8**	**0.8**
Total other pigs (b)	**118.2**	**126.3**	**110.0**	**105.9**	**110.3**
110kg and over	2.5	2.5	2.6	2.5	2.6
80kg and under 110kg	11.6	9.8	13.7	9.9	11.7
50kg and under 80kg	23.8	26.4	20.8	23.0	23.8
20kg and under 50kg	42.6	45.5	37.3	35.8	35.2
under 20kg	37.7	42.1	35.5	34.8	36.9
TOTAL SHEEP AND LAMBS	**8 416.3**	**8 721.2**	**9 000.8**	**9 129.7**	**9 461.6**
Total sheep and lambs one year old and over	**4 520.4**	**4 659.3**	**4 775.7**	**4 839.5**	**4 951.5**
Breeding flock - total	4 126.0	4 270.3	4 392.2	4 460.5	4 566.9
ewes kept for breeding	3 342.2	3 477.4	3 591.0	3 645.1	3 728.4
two-tooth ewes (shearling ewes/gimmers)	783.8	792.9	801.2	815.3	838.5
Other sheep one year old and over - total	394.5	389.0	383.5	379.0	384.6
rams kept for service	92.5	96.3	100.0	101.8	104.8
draft and cast ewes, wethers and others total	301.9	292.7	283.5	277.2	279.8
draft and cast ewes	231.8	228.4	221.0	215.8	219.7
wethers and other sheep	70.1	64.3	62.5	61.5	60.1
Lambs under one year old	**3 895.9**	**4 061.9**	**4 225.0**	**4 290.3**	**4 510.1**

Notes

(a) Includes estimates for minor holdings.

(b) Weights are liveweight.

Totals may not necessarily agree with the sum of their components due to rounding.

PIGS and SHEEP on Agricultural Holdings
WALES (a)
1987 - 1991 (at June Census)

thousands

1987	1988	1989	1990	1991	
129.9	**124.9**	**111.8**	**100.9**	**103.1**	**TOTAL PIGS**
18.5	**17.2**	**14.8**	**14.5**	**14.4**	**Total breeding pigs**
15.5	14.6	12.7	12.0	12.0	Breeding herd - total
11.9	10.7	9.5	9.2	9.0	mated sows and gilts - total
9.6	8.6	7.7	7.3	7.3	sows in pig
2.3	2.1	1.9	1.9	1.7	gilts in pig
3.7	3.9	3.1	2.9	2.9	other sows (being suckled or for breeding)
1.0	0.9	0.8	0.8	0.8	Boars for service
2.0	1.7	1.3	1.7	1.6	Gilts not yet in pig
0.7	**0.7**	**0.3**	**0.3**	**0.3**	**Barren sows for fattening**
110.6	**107.0**	**96.6**	**86.1**	**88.5**	**Total other pigs (b)**
1.9	1.8	1.8	0.9	1.2	110kg and over
10.3	9.7	9.6	9.0	9.3	80kg and under 110kg
22.2	22.9	17.9	18.5	18.1	50kg and under 80kg
38.2	36.1	33.2	28.4	30.1	20kg and under 50kg
38.0	36.6	34.0	29.3	29.8	under 20kg
9 795.8	**10 296.5**	**10 754.4**	**10 935.3**	**10 850.8**	**TOTAL SHEEP AND LAMBS**
5 100.8	**5 296.9**	**5 529.9**	**5 646.9**	**5 615.5**	**Total sheep and lambs one year old and over**
4 701.0	4 893.3	5 111.5	5 235.0	5 222.9	Breeding flock - total
3 822.3	3 961.7	4 107.7	4 261.7	4 313.2	ewes kept for breeding
878.6	931.6	1 003.8	973.3	909.8	two-tooth ewes (shearling ewes/gimmers)
399.8	403.6	418.4	412.0	392.6	Other sheep one year old and over - total
109.6	114.6	120.9	122.0	122.5	rams kept for service
290.2	289.0	297.6	290.0	270.1	draft and cast ewes, wethers and others total
225.9	230.2	233.3	231.7	214.9	draft and cast ewes
64.4	58.8	64.3	58.2	55.2	wethers and other sheep
4 695.0	**4 999.7**	**5 224.5**	**5 288.3**	**5 235.3**	**Lambs under one year old**

PIGS and SHEEP on Agricultural Holdings
SCOTLAND (a)
1982 - 1986 (at June Census)

Table 4.14

	1982	1983	1984	1985	1986
TOTAL PIGS	**461.0**	**442.2**	**405.5**	**418.5**	**413.7**
Total breeding pigs	**56.7**	**52.9**	**47.6**	**50.2**	**49.6**
Breeding herd - total	49.4	46.3	41.8	43.7	42.1
mated sows and gilts - total	38.9	36.2	33.3	35.2	33.6
sows in pig	31.9	30.0	27.2	29.0	27.8
gilts in pig	7.0	6.3	6.2	6.2	5.8
other sows (being suckled or for breeding)	10.5	10.1	8.5	8.4	8.5
Boars for service	2.5	2.5	2.2	2.4	2.3
Gilts not yet in pig	4.8	4.1	3.5	4.1	5.2
Barren sows for fattening	**0.6**	**1.2**	**0.6**	**0.6**	**0.7**
Total other pigs (b)	**403.7**	**388.1**	**357.4**	**367.7**	**363.4**
110kg and over	3.0	4.0	4.3	3.3	1.1
80kg and under 110kg	37.7	33.0	30.2	29.3	28.0
50kg and under 80kg	108.3	94.6	95.2	96.3	94.5
20kg and under 50kg	134.5	131.5	121.2	121.5	125.7
under 20kg	120.2	125.0	106.6	117.3	114.1
TOTAL SHEEP AND LAMBS	**7 960.6**	**8 024.0**	**8 145.4**	**8 358.4**	**8 514.9**
Total sheep and lambs one year old and over	**4 271.4**	**4 280.2**	**4 317.6**	**4 377.0**	**4 441.6**
Breeding flock - total	4 050.3	4 073.5	4 116.3	4 180.4	4 238.3
ewes kept for breeding	3 286.4	3 317.1	3 370.0	3 421.6	3 465.1
two-tooth ewes (shearling ewes/gimmers)	764.0	756.4	746.3	758.9	773.2
Other sheep one year old and over - total	221.1	206.6	201.3	196.5	203.3
rams kept for service	97.4	99.8	101.7	103.8	106.1
draft and cast ewes, wethers and others	123.7	106.8	99.6	92.8	97.2
Lambs under one year old	**3 689.2**	**3 743.9**	**3 827.7**	**3 981.5**	**4 073.3**

Notes

(a) Excludes estimates for minor holdings. In 1991 the Scottish census was revised to exclude returns from about 2,500 holdings (net) which were reclassified as minor holdings. Retrospective revisions on this basis have been made from 1987 to 1990.

(b) Weights are liveweight.

na not available.

Totals may not necessarily agree with the sum of their components due to rounding

PIGS and SHEEP on Agricultural Holdings
SCOTLAND (a)
1987 - 1991 (at June Census)

thousands

1987	1988	1989	1990	1991	
426.9	**467.4**	**462.5**	**449.4**	**493.0**	**TOTAL PIGS**
52.4	**56.0**	**54.4**	**60.6**	**64.9**	**Total breeding pigs**
44.6	47.7	47.1	51.0	51.4	Breeding herd - total
36.4	39.2	38.8	40.6	42.3	mated sows and gilts - total
30.0	32.8	32.7	33.8	33.5	sows in pig
6.4	6.5	6.0	6.8	8.8	gilts in pig
8.3	8.5	8.3	10.4	9.0	other sows (being suckled or for breeding)
2.5	2.7	2.7	3.1	3.1	Boars for service
5.3	5.6	4.6	6.5	10.4	Gilts not yet in pig
0.6	**0.5**	**0.4**	**0.8**	**0.5**	**Barren sows for fattening**
373.9	**411.0**	**407.7**	**388.0**	**427.6**	**Total other pigs (b)**
1.2	3.1	1.2	1.5	0.9	110kg and over
30.5	35.3	35.7	35.8	39.8	80kg and under 110kg
99.9	112.5	108.9	97.9	117.9	50kg and under 80kg
121.9	133.9	137.4	128.1	141.9	20kg and under 50kg
120.3	126.1	124.5	124.7	127.2	under 20kg
8 857.9	**9 249.5**	**9 376.0**	**9 582.7**	**9 757.4**	**TOTAL SHEEP AND LAMBS**
4 569.8	**4 742.2**	**4 803.9**	**4 830.9**	**4 914.4**	**Total sheep and lambs one year old and over**
4 384.6	4 531.0	4 601.2	4 632.4	4 692.5	Breeding flock - total
3 607.3	3 717.2	3 778.9	3 814.0	3 863.9	ewes kept for breeding
777.3	813.9	822.3	818.4	828.6	two-tooth ewes (shearling ewes/gimmers)
185.1	211.2	202.8	198.5	222.0	Other sheep one year old and over - total
105.6	110.1	112.5	113.7	116.3	rams kept for service
79.5	101.1	90.3	84.8	105.6	draft and cast ewes, wethers and others
4 288.1	**4 507.3**	**4 572.1**	**4 751.8**	**4 843.0**	**Lambs under one year old**

PIGS and SHEEP on Agricultural Holdings
NORTHERN IRELAND (a)
1982 - 1986 (at June Census)

Table 4.15

	1982	1983	1984	1985	1986
TOTAL PIGS	**640.2**	**650.9**	**615.1**	**616.9**	**617.4**
Total breeding pigs	**79.1**	**77.0**	**70.4**	**72.0**	**70.4**
Breeding herd - total	70.4	69.5	63.0	64.4	62.5
mated sows and gilts - total	50.6	48.7	45.3	46.5	44.9
sows in pig	40.2	41.0	38.3	38.6	37.4
gilts in pig	10.4	7.6	7.0	7.9	7.5
other sows (being suckled or for breeding)	19.8	20.8	17.6	18.0	17.6
Boars for service	3.4	3.4	3.1	3.2	3.1
Gilts not yet in pig	5.3	4.1	4.3	4.4	4.8
Barren sows for fattening (b)	*na*	*na*	*na*	*na*	*na*
Total other pigs (c)	**561.1**	**574.0**	**544.7**	**544.9**	**547.0**
110kg and over (b)	8.2	9.6	8.9	8.9	9.4
80kg and under 110kg	53.6	55.7	53.3	52.6	49.2
50kg and under 80kg	168.0	169.0	162.1	159.5	169.4
20kg and under 50kg	176.6	182.0	173.4	170.4	167.8
under 20kg	154.7	157.7	147.0	153.4	151.2
TOTAL SHEEP AND LAMBS	**1 233.8**	**1 323.5**	**1 449.6**	**1 590.5**	**1 700.5**
Total sheep and lambs one year old and over	**694.3**	**735.3**	**776.8**	**838.9**	**895.2**
Breeding flock - total (d)	626.1	670.4	708.6	769.1	818.8
ewes kept for breeding (e)	516.2	552.2	585.5	628.8	671.1
two-tooth ewes (shearling ewes/gimmers) (f)	109.9	118.2	123.1	140.3	147.7
Other sheep one year old and over - total	68.2	64.9	68.2	69.8	76.4
rams kept for service	16.5	17.8	18.8	20.4	21.5
draft and cast ewes, wethers and others	51.8	47.2	49.4	49.3	55.0
Lambs under one year old	**539.4**	**588.2**	**672.8**	**751.6**	**805.3**

Notes

(a) Excludes estimates for minor holdings.

(b) Barren sows for fattening are included in other pigs 110 kg and over.

(c) Weights are liveweight.

(d) Collected as "Ewes to be put to the ram this year".

(e) Collected as "1 to 2 years old".

(f) Collected as "2 years old and over".

na not available.

Totals may not necessarily agree with the sum of their components due to rounding.

PIGS and SHEEP on Agricultural Holdings
NORTHERN IRELAND (a)
1987 - 1991 (at June Census)

thousands

1987	1988	1989	1990	1991	
607.0	**619.1**	**588.8**	**590.4**	**588.3**	**TOTAL PIGS**
67.4	**69.6**	**65.0**	**65.6**	**65.9**	**Total breeding pigs**
60.3	62.3	58.2	58.7	59.1	Breeding herd - total
43.4	45.1	42.4	43.8	43.8	mated sows and gilts - total
36.5	37.8	35.6	36.2	36.5	sows in pig
6.9	7.3	6.8	7.6	7.3	gilts in pig
16.9	17.2	15.9	14.9	15.3	other sows (being suckled or for breeding)
3.0	3.0	2.9	2.9	2.9	Boars for service
4.1	4.3	3.8	4.0	3.9	Gilts not yet in pig
na	*na*	*na*	*na*	*na*	**Barren sows for fattening (b)**
539.5	**549.5**	**523.8**	**524.8**	**522.5**	**Total other pigs (c)**
7.6	7.3	4.7	4.6	4.2	110kg and over (b)
54.9	57.5	64.3	62.3	58.4	80kg and under 110kg
165.1	165.9	155.0	151.1	155.4	50kg and under 80kg
167.2	166.6	158.9	167.3	161.5	20kg and under 50kg
144.7	152.2	140.9	139.5	143.0	under 20kg
1 857.7	**2 073.1**	**2 316.7**	**2 504.9**	**2 574.2**	**TOTAL SHEEP AND LAMBS**
964.5	**1 070.2**	**1 193.0**	**1 281.4**	**1 314.8**	**Total sheep and lambs one year old and over**
885.6	979.7	1 085.1	1 168.4	1 203.8	Breeding flock - total (d)
724.5	785.5	861.4	938.9	984.7	ewes kept for breeding (e)
161.1	194.2	223.7	229.5	219.1	two-tooth ewes (shearling ewes/gimmers) (f)
78.9	90.5	107.9	113.0	111.0	Other sheep one year old and over - total
23.0	25.7	28.9	30.8	32.0	rams kept for service
55.9	64.8	79.0	82.2	79.0	draft and cast ewes, wethers and others
893.2	**1 002.9**	**1 123.7**	**1 223.5**	**1 259.4**	**Lambs under one year old**

POULTRY on Agricultural Holdings
UNITED KINGDOM (a)
1982 - 1986 (at June Census)

Table 4.16

	1982	1983	1984	1985	1986
Total fowls	**126 091**	**117 854**	**118 846**	**119 456**	**120 740**
Growing pullets (day old to point of lay)	**14 766**	**11 828**	**12 536**	**12 503**	**12 502**
Total laying flock	**44 792**	**41 127**	**40 573**	**39 538**	**38 096**
in flock for: less than 12 months	32 711	29 930	29 159	29 147	29 778
12 months or more	12 081	11 197	11 414	10 391	8 318
Total breeding flock	**6 457**	**6 012**	**6 396**	**6 104**	**6 334**
Breeding hens (b)	5 213	4 844	5 198	5 019	5 187
Cocks and cockerels for breeding (b)	579	565	615	569	569
Table fowls	**60 075**	**58 887**	**59 341**	**61 311**	**63 807**
Total ducks and geese (c)	**na**	**1 566**	**1 530**	**1 654**	**1 756**
Ducks (d)	1 443	1 410	1 367	1 484	1 590
Geese (d)	157	138	146	154	150

Notes

(a) Estimates for minor holdings are included for England and Wales but not for Scotland and Northern Ireland. In 1991 the Scottish census was revised to exclude returns from about 2,500 holdings (net) which were reclassified as minor holdings. Retrospective revisions on this basis have been made from 1987 to 1990.

(b) Great Britain only. Not collected separately in Northern Ireland.

(c) Includes guinea fowl in Scotland.

(d) Excluding Scotland.

na not available.

Totals may not necessarily agree with the sum of their components due to rounding.

POULTRY on Agricultural Holdings
UNITED KINGDOM (a)
1987 - 1991 (at June Census)

thousands

1987	1988	1989	1990	1991	
128 801	**130 998**	**120 351**	**124 615**	**127 228**	**Total fowls**
12 238	**11 243**	**9 414**	**10 452**	**11 016**	**Growing pullets (day old to point of lay)**
38 529	**37 420**	**33 957**	**33 468**	**33 273**	**Total laying flock**
28 557	28 633	26 388	26 528	26 260	in flock for: less than 12 months
9 972	8 787	7 569	6 940	7 013	12 months or more
7 165	**6 898**	**6 805**	**7 107**	**7 238**	**Total breeding flock**
5 941	5 631	5 483	5 592	5 715	Breeding hens (b)
622	608	563	601	559	Cocks and cockerels for breeding (b)
70 869	**75 437**	**70 176**	**73 588**	**75 701**	**Table fowls**
1 768	**1 848**	**2 110**	**2 217**	**2 191**	**Total ducks and geese (c)**
1 562	1 635	1 884	1 967	1 987	Ducks (d)
178	182	196	158	146	Geese (d)

POULTRY on Agricultural Holdings
ENGLAND (a)
1982 - 1986 (at June Census)

Table 4.17

	1982	1983	1984	1985	1986
Total fowls	**95 150**	**89 671**	**89 228**	**89 868**	**91 960**
Growing pullets (day old to point of lay)	**11 566**	**9 364**	**10 065**	**10 008**	**10 119**
Total laying flock	**34 900**	**32 804**	**32 556**	**31 881**	**30 924**
in flock for: less than 12 months	24 601	23 197	23 034	22 805	23 939
12 to 18 months	8 381	7 803	7 668	7 468	5 702
18 months or more	1 917	1 804	1 854	1 608	1 284
Total breeding flock	**4 345**	**4 033**	**4 160**	**3 916**	**4 045**
Breeding hens total	3 907	3 608	3 738	3 525	3 666
laying eggs to hatch: layer chicks	*na*	*na*	*na*	*na*	*na*
table chicks	*na*	*na*	*na*	*na*	*na*
Cocks and cockerels for breeding	438	425	422	390	378
Table fowls	**44 340**	**43 469**	**42 447**	**44 063**	**46 872**
Total ducks and geese	**1 538**	**1 494**	**1 454**	**1 575**	**1 673**
Ducks	1 409	1 377	1 331	1 456	1 542
Geese	130	117	123	119	131

Notes

(a)   Estimates for minor holdings are included.

na   not available.

Totals may not necessarily agree with the sum of their components due to rounding.

POULTRY on Agricultural Holdings
ENGLAND (a)
1987 - 1991 (at June Census)

thousands

1987	1988	1989	1990	1991	
97 874	**98 862**	**90 387**	**93 121**	**95 715**	**Total fowls**
9 728	**8 958**	**7 600**	**8 376**	**8 806**	**Growing pullets (day old to point lay)**
31 290	**30 199**	**27 596**	**26 997**	**26 892**	**Total laying flock**
22 519	22 525	21 163	20 846	20 538	in flock for: less than 12 months
7 307	6 296	5 198	4 931	5 319	12 to 18 months
1 463	1 379	1 235	1 219	1 034	18 months or more
4 603	**4 355**	**4 363**	**4 224**	**4 410**	**Total breeding flock**
4 186	3 946	3 978	3 814	4 006	Breeding hens total
na	1 011	653	684	732	laying eggs to hatch: layer chicks
na	2 935	3 326	3 130	3 275	table chicks
417	409	385	410	404	Cocks and cockerels for breeding
52 254	**55 350**	**50 827**	**53 524**	**55 608**	**Table fowls**
1 666	**1 733**	**1 998**	**2 096**	**2 097**	**Total ducks and geese**
1 508	1 571	1 822	1 942	1 955	Ducks
158	162	176	154	141	Geese

POULTRY on Agricultural Holdings
WALES (a)
1982 - 1986 (at June Census)

Table 4.18

	1982	1983	1984	1985	1986
Total fowls	**7 611**	**5 701**	**6 418**	**6 672**	**6 678**
Growing pullets (day old to point of lay)	**771**	**480**	**414**	**475**	**414**
Total laying flock	**2 054**	**1 392**	**1 320**	**1 195**	**1 229**
in flock for: less than 12 months	1 540	992	849	803	775
12 to 18 months	394	282	267	282	345
18 months or more	121	117	204	110	109
Total breeding flock	**370**	**255**	**448**	**442**	**454**
Breeding hens total	335	227	407	395	409
laying eggs to hatch: layer chicks	*na*	*na*	*na*	*na*	112
table chicks	*na*	*na*	*na*	*na*	297
Cocks and cockerels for breeding	35	28	41	47	45
Table fowls	**4 415**	**3 574**	**4 237**	**4 559**	**4 581**
Total ducks and geese	**40**	**35**	**34**	**33**	**39**
Ducks	22	20	19	20	26
Geese	18	14	14	13	13

Notes
(a) Estimates for minor holdings are included.
na not available.
Totals may not necessarily agree with the sum of their components due to rounding.

POULTRY on Agricultural Holdings
WALES (a)
1987 - 1991 (at June Census)

thousands

1987	1988	1989	1990	1991	
6 833	**6 986**	**6 767**	**6 709**	**6 907**	**Total fowls**
429	**527**	**420**	**515**	**516**	**Growing pullets (day old to point of lay)**
1 161	**1 169**	**1 106**	**1 072**	**1 038**	**Total laying flock**
816	860	730	704	739	in flock for: less than 12 months
251	220	237	223	213	12 to 18 months
93	90	139	145	87	18 months or more
536	**461**	**361**	**418**	**660**	**Total breeding flock**
480	408	326	375	620	Breeding hens total
174	6	3	2	36	laying eggs to hatch: layer chicks
306	401	323	372	583	table chicks
55	53	36	43	40	Cocks and cockerels for breeding
4 707	**4 829**	**4 880**	**4 704**	**4 694**	**Table fowls**
48	**60**	**56**	**60**	**26**	**Total ducks and geese**
34	45	42	*na*	*na*	Ducks
14	14	15	*na*	*na*	Geese

POULTRY on Agricultural Holdings
SCOTLAND (a)
1982 - 1986 (at June Census)

Table 4.19

	1982	1983	1984	1985	1986
Total fowls	**12 498**	**12 272**	**12 622**	**13 071**	**12 402**
Growing pullets (day old to point of lay)	**1 145**	**1 026**	**974**	**1 093**	**1 013**
Total laying flock	**3 560**	**3 172**	**3 290**	**3 283**	**3 010**
in flock for:　less than 12 months	2 896	2 654	2 773	2 891	2 684
12 months or more	664	518	516	393	326
Total breeding flock	**1 077**	**1 121**	**1 206**	**1 230**	**1 257**
Breeding hens	970	1 009	1 054	1 099	1 111
Cocks and cockerels for breeding	107	112	152	132	146
Table fowls	**6 717**	**6 954**	**7 153**	**7 464**	**7 122**
Total ducks and geese (b)	*na*	**17**	**17**	**16**	**16**
Ducks	*na*	*na*	*na*	*na*	*na*
Geese	*na*	*na*	*na*	*na*	*na*

Notes
(a) Excludes estimates for minor holdings. In 1991 the Scottish census was revised to exclude returns from about 2,500 holdings (net)
which were reclassified as minor holdings. Retrospective revisions on this basis have been made from 1987 to 1990.
(b) Including guinea fowl.
na not available.
Totals may not necessarily agree with the sum of their components due to rounding.

POULTRY on Agricultural Holdings
SCOTLAND (a)
1987 - 1991 (at June Census)

thousands

1987	1988	1989	1990	1991	
13 982	15 091	13 667	14 670	13 605	**Total fowls**
1 063	919	605	724	748	**Growing pullets (day old to point of lay)**
2 907	2 862	2 432	2 533	2 595	**Total laying flock**
2 592	2 578	2 229	2 385	2 466	in flock for: less than 12 months
315	284	203	148	129	12 months or more
1 425	1 422	1 322	1 552	1 204	**Total breeding flock**
1 275	1 277	1 179	1 403	1 089	Breeding hens
150	145	142	149	116	Cocks and cockerels for breeding
8 587	9 887	9 309	9 862	9 057	**Table fowls**
29	31	31	31	31	**Total ducks and geese (b)**
na	na	na	na	na	Ducks
na	na	na	na	na	Geese

POULTRY on Agricultural Holdings
NORTHERN IRELAND (a)
1982 - 1986 (at June Census)

Table 4.20

	1982	1983	1984	1985	1986
Total fowls	**10 832**	**10 210**	**10 577**	**9 845**	**9 700**
Growing pullets (day old to point of lay)	**1 285**	**957**	**1 083**	**927**	**957**
Total laying flock	**4 279**	**3 759**	**3 408**	**3 178**	**2 933**
in flock for: less than 12 months (b)	3 674	3 087	2 502	2 648	2 380
12 months or more (c)	605	672	905	530	553
Total breeding flock	**665**	**603**	**582**	**516**	**578**
Breeding hens	*na*	*na*	*na*	*na*	*na*
Cocks and cockerels for breeding	*na*	*na*	*na*	*na*	*na*
Table fowls	**4 603**	**4 890**	**5 505**	**5 224**	**5 232**
Total ducks and geese	**22**	**20**	**26**	**30**	**28**
Ducks	13	13	16	22	22
Geese	10	7	9	8	6

Notes

(a) Excludes estimates for minor holdings.

(b) Collected as "Pullets (from point of lay to end of first laying cycle)".

(c) Collected as "Hens (retained after first laying cycle)".

na not available.

Totals may not necessarily agree with the sum of their components due to rounding.

POULTRY on Agricultural Holdings
NORTHERN IRELAND (a)
1987 - 1991 (at June Census)

thousands

1987	1988	1989	1990	1991	
10 113	**10 059**	**9 530**	**10 114**	**11 000**	**Total fowls**
1 019	**839**	**789**	**836**	**946**	**Growing pullets (day old to point of lay)**
3 171	**3 189**	**2 823**	**2 866**	**2 748**	**Total laying flock**
2 630	2 670	2 265	2 593	2 518	in flock for: less than 12 months (b)
542	519	557	273	230	12 months or more (c)
601	**659**	**759**	**913**	**964**	**Total breeding flock**
na	*na*	*na*	*na*	*na*	Breeding hens
na	*na*	*na*	*na*	*na*	Cocks and cockerels for breeding
5 322	**5 372**	**5 160**	**5 499**	**6 342**	**Table fowls**
25	**24**	**26**	**30**	**37**	**Total ducks and geese**
19	18	20	25	32	Ducks
6	6	6	5	5	Geese

LABOUR FORCE on Agricultural Holdings
UNITED KINGDOM (a)
1982 - 1986 (at June Census)

Table 4.21

	1982	1983	1984	1985	1986
TOTAL LABOUR FORCE	**705.6**	**699.3**	**691.2**	**690.6**	**679.9**
Total farmers,partners,directors (doing farm work) (b)	**292.8**	**289.6**	**292.2**	**290.3**	**288.9**
Whole-time - total	204.0	202.8	202.5	199.3	197.4
principal farmers and partners (c)	161.1	159.9	159.5	156.9	155.0
other partners and directors (d)	42.9	42.8	43.0	42.4	42.4
Part-time - total	88.7	86.8	89.7	91.0	91.5
principal farmers and partners (c)	66.2	64.1	66.8	69.2	68.9
other partners and directors (d)	22.6	22.7	22.9	21.8	22.6
Spouses of farmers,partners,directors	**74.0**	**75.7**	**74.9**	**76.9**	**77.1**
Salaried managers (d)	**7.9**	**7.8**	**7.8**	**8.3**	**8.3**
Total other workers (e)	**331.0**	**326.2**	**316.3**	**315.2**	**305.5**
Male	243.8	240.3	233.0	231.1	223.5
Female	87.2	85.9	83.4	84.1	82.1
Regular family workers - total	55.0	54.3	53.6	55.5	57.7
whole-time - total	35.3	35.0	34.7	35.5	37.0
male	30.1	30.0	29.8	30.7	32.3
female	5.2	5.0	4.8	4.7	4.7
part-time - total	19.7	19.3	18.9	20.0	20.7
male	12.6	12.5	12.1	12.8	13.4
female	7.1	6.8	6.8	7.2	7.3
Regular hired workers - total	177.2	174.0	167.2	162.5	152.6
whole-time - total	135.3	132.7	126.1	121.2	111.9
male	124.4	122.2	115.7	110.8	101.8
female	10.8	10.5	10.4	10.4	10.1
part-time - total	41.9	41.3	41.1	41.3	40.7
male	19.3	18.8	18.6	18.6	18.8
female	22.6	22.5	22.5	22.7	21.9
Seasonal or casual workers - total	98.7	97.9	95.6	97.2	95.3
male	57.3	56.9	56.7	58.1	57.2
female	41.5	41.0	38.8	39.1	38.1

Notes

(a) Estimates for minor holdings are included for England and Wales but not for Scotland and Northern Ireland. In 1991 the Scottish census was revised to exclude returns from about 2,500 holdings (net) which were reclassified as minor holdings. Retrospective revisions on this basis have been made from 1987 to 1990.

(b) Figures exclude the wives/husbands of farmers, partners and directors, even though they themselves may be partners or directors.

(c) Includes other partners and directors in Northern Ireland.

(d) Great Britain only.

LABOUR FORCE on Agricultural Holdings
UNITED KINGDOM (a)
1987 - 1991 (at June Census)

thousands

1987	1988	1989	1990	1991	
665.1	**657.2**	**643.7**	**642.0**	**627.9**	**TOTAL LABOUR FORCE**
284.5	**284.2**	**283.3**	**281.6**	**278.6**	**Total farmers,partners,directors (doing farm work) (b)**
194.3	192.5	188.9	183.5	177.7	Whole-time - total
153.2	152.2	149.2	144.4	140.1	principal farmers and partners (c)
41.2	40.3	39.7	39.1	37.6	other partners and directors (d)
90.1	91.6	94.4	98.1	100.9	Part-time - total
67.4	69.1	70.6	73.7	76.5	principal farmers and partners (c)
22.7	22.5	23.8	24.4	24.4	other partners and directors (d)
77.1	**76.8**	**76.4**	**77.1**	**76.5**	**Spouses of farmers,partners,directors**
7.9	**7.9**	**7.8**	**8.1**	**7.9**	**Salaried managers (d)**
295.7	**288.4**	**276.3**	**275.3**	**264.9**	**Total other workers (e)**
214.7	208.2	199.2	196.7	189.6	Male
81.0	80.2	77.1	78.6	75.3	Female
54.0	51.7	49.4	48.6	47.9	Regular family workers - total
33.9	31.8	30.3	28.9	27.9	whole-time - total
29.6	27.6	26.2	25.0	24.2	male
4.3	4.2	4.1	3.9	3.7	female
20.1	19.9	19.1	19.7	19.9	part-time - total
13.1	12.7	12.2	12.7	12.9	male
7.0	7.2	6.8	6.9	7.1	female
148.2	143.9	138.7	136.2	130.4	Regular hired workers - total
108.0	103.4	99.4	96.3	91.6	whole-time - total
97.8	93.1	88.4	84.7	80.4	male
10.2	10.3	11.0	11.6	11.2	female
40.1	40.5	39.3	39.9	38.8	part-time - total
18.3	18.6	18.4	18.7	18.3	male
21.8	21.8	20.9	21.2	20.5	female
93.5	92.8	88.3	90.5	86.6	Seasonal or casual workers - total
55.9	56.2	54.0	55.6	53.8	male
37.7	36.7	34.3	34.9	32.8	female

(e) In England and Wales figures exclude school children but include trainees employed under an official youth training scheme
   at Agricultural Board rates or above. In Scotland and Northern Ireland school children and all trainees are excluded.
Totals may not necessarily agree with the sum of their components due to rounding.

Table 4.22

	1982	1983	1984	1985	1986
TOTAL LABOUR FORCE	**510.0**	**505.2**	**497.7**	**495.3**	**488.5**
Total farmers,partners,directors (doing farm work)	**192.1**	**189.8**	**190.1**	**187.8**	**187.4**
Whole-time - total	133.6	132.9	131.7	129.8	128.5
principal farmers and partners	100.9	100.3	99.2	97.9	96.8
other partners and directors	32.7	32.6	32.4	31.9	31.7
Part-time - total	58.5	56.9	58.5	58.0	58.9
principal farmers and partners	39.3	38.0	39.0	39.5	39.8
other partners and directors	19.2	18.9	19.4	18.4	19.2
Total spouses of farmers,partners,directors	**50.1**	**50.7**	**50.5**	**51.6**	**51.5**
Spouses of principal farmers and partners	46.0	46.8	46.6	47.3	47.3
Spouses of other partners and directors	4.1	3.9	4.0	4.3	4.2
Salaried managers	**6.7**	**6.6**	**6.7**	**7.1**	**7.1**
Total other workers (b)	**261.2**	**258.1**	**250.4**	**248.9**	**242.4**
Male	185.7	183.3	177.5	175.3	169.9
Female	75.5	74.8	72.9	73.6	72.5
Regular family workers - total	31.6	31.0	31.0	32.6	35.0
whole-time - total	19.8	19.7	19.8	20.7	22.5
male	17.2	17.1	17.3	18.1	19.9
female	2.6	2.6	2.6	2.6	2.6
part-time - total	11.7	11.3	11.2	11.8	12.5
male	6.8	6.7	6.4	6.9	7.4
female	4.9	4.7	4.7	5.0	5.2
Regular hired workers - total	150.5	148.7	143.6	139.6	131.7
whole-time - total	114.6	113.0	108.0	104.0	96.6
male	104.9	103.5	98.5	94.5	87.2
female	9.7	9.5	9.5	9.5	9.4
part-time - total	35.9	35.6	35.6	35.6	35.1
male	15.2	14.9	14.7	14.6	14.8
female	20.7	20.8	20.9	21.0	20.3
Seasonal or casual workers - total	79.1	78.4	75.8	76.7	75.7
male	41.5	41.2	40.6	41.2	40.7
female	37.6	37.2	35.2	35.5	35.0

Notes

(a) Estimates for minor holdings are included.

(b) Figures exclude school children but include trainees employed under an official youth training scheme at Agricultural Wages Board rates or above.

Totals may not necessarily agree with the sum of their components due to rounding.

LABOUR FORCE on Agricultural Holdings
ENGLAND (a)
1987 - 1991 (at June Census)

thousands

1987	1988	1989	1990	1991	
477.9	**469.9**	**457.4**	**455.9**	**445.0**	**TOTAL LABOUR FORCE**
185.5	**183.5**	**182.4**	**181.9**	**180.3**	**Total farmers,partners,directors (doing farm work)**
126.7	124.5	122.4	119.0	115.0	Whole-time - total
95.7	94.1	92.5	89.8	87.0	principal farmers and partners
31.0	30.4	29.9	29.2	28.0	other partners and directors
58.8	59.0	60.1	63.0	65.3	Part-time - total
39.6	39.8	39.9	42.6	44.7	principal farmers and partners
19.2	19.3	20.2	20.4	20.7	other partners and directors
51.2	**51.2**	**50.7**	**50.5**	**49.6**	**Total spouses of farmers,partners,directors**
47.1	47.1	46.6	46.5	45.7	Spouses of principal farmers and partners
4.2	4.1	4.0	4.0	4.0	Spouses of other partners and directors
7.0	**7.0**	**6.9**	**7.2**	**7.1**	**Salaried managers**
234.2	**228.2**	**217.4**	**216.2**	**207.9**	**Total other workers (b)**
162.7	157.8	150.1	147.6	142.7	Male
71.5	70.3	67.3	68.6	65.3	Female
34.1	32.7	30.8	29.9	29.9	Regular family workers - total
21.8	20.4	19.4	18.5	18.1	whole-time - total
19.2	18.0	17.1	16.3	16.0	male
2.6	2.4	2.3	2.2	2.1	female
12.3	12.3	11.3	11.4	11.8	part-time - total
7.3	7.2	6.7	6.8	7.1	male
5.1	5.1	4.6	4.6	4.7	female
126.0	122.4	118.0	115.8	110.9	Regular hired workers - total
91.5	87.3	84.1	81.5	77.5	whole-time - total
82.1	78.0	73.9	70.7	67.2	male
9.3	9.3	10.2	10.8	10.3	female
34.5	35.1	33.9	34.3	33.3	part-time - total
14.4	14.9	14.7	14.8	14.6	male
20.1	20.2	19.2	19.4	18.8	female
74.1	73.1	68.6	70.6	67.2	Seasonal or casual workers - total
39.6	39.8	37.7	39.0	37.8	male
34.4	33.2	30.9	31.6	29.4	female

LABOUR FORCE on Agricultural Holdings
WALES (a)
1982 - 1986 (at June Census)

Table 4.23

	1982	1983	1984	1985	1986
TOTAL LABOUR FORCE	**66.5**	**65.3**	**66.8**	**68.4**	**67.7**
Total farmers,partners,directors (doing farm work)	**34.6**	**33.2**	**35.7**	**36.3**	**35.7**
Whole-time - total	24.6	24.2	25.4	24.5	24.4
principal farmers and partners	19.7	19.6	20.5	19.8	19.6
other partners and directors	5.0	4.6	4.8	4.7	4.8
Part-time - total	10.0	9.0	10.3	11.9	11.3
principal farmers and partners	7.2	6.0	7.5	9.2	8.6
other partners and directors	2.8	3.1	2.8	2.7	2.8
Total spouses of farmers,partners,directors	**10.7**	**11.6**	**11.2**	**11.5**	**11.6**
Spouses of principal farmers and partners	10.1	11.0	10.6	10.9	10.9
Spouses of other partners and directors	0.6	0.6	0.6	0.6	0.6
Salaried managers	**0.3**	**0.3**	**0.3**	**0.3**	**0.3**
Total other workers (b)	**20.9**	**20.3**	**19.7**	**20.4**	**20.1**
Male	16.7	16.3	16.1	16.8	16.5
Female	4.2	4.0	3.6	3.6	3.5
Regular family workers - total	5.3	5.3	4.9	5.3	5.6
whole-time - total	3.4	3.4	3.1	3.3	3.5
male	2.7	2.7	2.5	2.7	2.9
female	0.7	0.6	0.6	0.6	0.6
part-time - total	1.9	1.9	1.8	2.0	2.1
male	1.3	1.3	1.2	1.4	1.4
female	0.6	0.6	0.6	0.6	0.7
Regular hired workers - total	6.6	6.4	5.9	5.9	5.7
whole-time - total	4.8	4.6	4.2	4.0	3.7
male	4.4	4.3	3.9	3.7	3.5
female	0.4	0.4	0.3	0.3	0.3
part-time - total	1.7	1.7	1.7	1.8	2.0
male	1.3	1.3	1.3	1.4	1.5
female	0.5	0.5	0.4	0.4	0.5
Seasonal or casual workers - total	9.0	8.7	8.9	9.2	8.8
male	7.0	6.8	7.2	7.5	7.3
female	2.0	1.9	1.7	1.7	1.5

Notes

(a) Estimates for minor holdings are included.

(b) Figures exclude school children but include trainees employed under an official youth training scheme at Agricultural Wages Board rates or above.

Totals may not necessarily agree with the sum of their components due to rounding.

LABOUR FORCE on Agricultural Holdings
WALES (a)
1987 - 1991 (at June Census)

thousands

1987	1988	1989	1990	1991	
65.6	**65.8**	**65.6**	**65.0**	**64.6**	**TOTAL LABOUR FORCE**
33.8	**34.9**	**35.3**	**35.0**	**34.2**	**Total farmers,partners,directors (doing farm work)**
23.5	23.8	23.8	23.4	22.6	Whole-time - total
19.2	19.5	19.4	18.9	18.5	principal farmers and partners
4.4	4.4	4.4	4.5	4.1	other partners and directors
10.3	11.0	11.5	11.6	11.7	Part-time - total
7.6	8.6	8.7	8.5	8.8	principal farmers and partners
2.7	2.5	2.8	3.2	2.9	other partners and directors
11.8	**11.9**	**12.0**	**11.6**	**11.9**	**Total spouses of farmers,partners,directors**
11.2	11.3	11.4	11.1	11.4	Spouses of principal farmers and partners
0.6	0.6	0.6	0.6	0.5	Spouses of other partners and directors
0.3	**0.3**	**0.3**	**0.3**	**0.3**	**Salaried managers**
19.7	**18.8**	**18.0**	**18.1**	**18.1**	**Total other workers (b)**
16.1	15.3	14.7	14.7	14.6	Male
3.5	3.6	3.3	3.4	3.6	Female
5.5	5.1	4.8	4.9	4.8	Regular family workers - total
3.4	3.0	2.9	2.8	2.7	whole-time - total
2.8	2.6	2.4	2.3	2.3	male
0.5	0.5	0.4	0.5	0.4	female
2.1	2.0	1.9	2.1	2.2	part-time - total
1.5	1.4	1.3	1.5	1.5	male
0.6	0.6	0.6	0.7	0.7	female
5.3	5.1	5.0	5.0	4.8	Regular hired workers - total
3.5	3.4	3.2	3.1	3.0	whole-time - total
3.2	3.1	3.0	2.8	2.7	male
0.3	0.3	0.3	0.3	0.3	female
1.8	1.8	1.8	1.9	1.8	part-time - total
1.4	1.4	1.4	1.4	1.4	male
0.4	0.4	0.4	0.4	0.4	female
8.8	8.6	8.3	8.2	8.5	Seasonal or casual workers - total
7.2	6.8	6.6	6.7	6.8	male
1.7	1.8	1.6	1.5	1.7	female

LABOUR FORCE on Agricultural Holdings
SCOTLAND (a)
1982 - 1986 (at June Census)

Table 4.24

	1982	1983	1984	1985	1986
TOTAL LABOUR FORCE	**68.8**	**67.9**	**66.2**	**66.2**	**63.8**
Total farmers,partners,directors (doing farm work) (b)	**29.4**	**29.8**	**29.8**	**30.1**	**29.9**
Whole-time - total	19.5	19.7	19.5	19.7	19.4
principal farmers and partners	14.2	14.1	13.8	13.9	13.5
other partners and directors	5.3	5.6	5.7	5.8	5.9
Part-time - total	9.8	10.1	10.3	10.4	10.5
principal farmers and partners	9.2	9.3	9.5	9.8	9.9
other partners and directors	0.6	0.8	0.8	0.7	0.7
Spouses of farmers,partners,directors	**8.0**	**8.4**	**8.3**	**8.5**	**8.6**
Salaried managers	**1.0**	**0.9**	**0.9**	**0.9**	**0.9**
Total other workers (c)	**30.5**	**28.9**	**27.2**	**26.6**	**24.3**
Male	26.0	24.7	23.3	22.6	21.0
Female	4.4	4.2	3.9	4.0	3.4
Regular family workers - total	9.6	9.7	9.5	9.5	9.3
whole-time - total	7.5	7.5	7.4	7.3	7.2
male	6.7	6.8	6.7	6.7	6.6
female	0.7	0.7	0.7	0.6	0.6
part-time - total	2.2	2.2	2.1	2.2	2.2
male	1.4	1.3	1.3	1.3	1.4
female	0.8	0.9	0.8	0.9	0.8
Regular hired workers - total	16.1	15.1	14.0	13.5	11.7
whole-time - total	13.3	12.6	11.6	11.0	9.5
male	12.8	12.2	11.2	10.6	9.2
female	0.5	0.4	0.4	0.4	0.3
part-time - total	2.8	2.5	2.4	2.5	2.3
male	1.6	1.4	1.4	1.5	1.3
female	1.2	1.1	1.0	1.0	0.9
Seasonal or casual workers - total	4.7	4.1	3.7	3.6	3.3
male	3.5	3.0	2.6	2.5	2.5
female	1.2	1.1	1.0	1.1	0.8

Notes

(a) Excludes estimates for minor holdings. In 1991 the Scottish census was revised to exclude returns from about 2,500 holdings (net) which were reclassified as minor holdings. Retrospective revisions on this basis have been made from 1987 to 1990.

(b) Figures exclude the wives/husbands of farmers, partners and directors, even though they themselves may be partners or directors.

(c) Excludes school children and all trainees.

Totals may not necessarily agree with the sum of their components due to rounding.

LABOUR FORCE on Agricultural Holdings
SCOTLAND (a)
1987 - 1991 (at June Census)

thousands

1987	1988	1989	1990	1991	
62.1	**62.0**	**61.6**	**62.7**	**61.1**	**TOTAL LABOUR FORCE**
29.7	**30.0**	**30.5**	**30.0**	**29.8**	**Total farmers,partners,directors (doing farm work) (b)**
19.5	19.6	19.5	18.4	17.8	Whole-time - total
13.8	14.1	14.1	13.0	12.3	principal farmers and partners
5.8	5.5	5.4	5.3	5.5	other partners and directors
10.2	10.5	11.0	11.6	11.9	Part-time - total
9.3	9.7	10.2	10.8	11.1	principal farmers and partners
0.8	0.8	0.8	0.8	0.9	other partners and directors
8.7	**8.7**	**8.9**	**10.2**	**10.4**	**Spouses of farmers,partners,directors**
0.6	**0.6**	**0.6**	**0.6**	**0.5**	**Salaried managers**
23.1	**22.6**	**21.7**	**21.9**	**20.4**	**Total other workers (c)**
19.8	19.4	18.5	18.6	17.2	Male
3.3	3.3	3.1	3.3	3.2	Female
6.8	6.6	6.4	6.6	6.3	Regular family workers - total
5.0	4.9	4.6	4.6	4.3	whole-time - total
4.6	4.5	4.2	4.3	4.0	male
0.4	0.4	0.3	0.4	0.3	female
1.8	1.8	1.8	2.0	2.0	part-time - total
1.1	1.1	1.2	1.3	1.2	male
0.7	0.6	0.6	0.7	0.7	female
13.4	13.0	12.3	12.0	11.4	Regular hired workers - total
11.0	10.7	10.1	9.6	9.0	whole-time - total
10.5	10.2	9.7	9.3	8.6	male
0.5	0.5	0.4	0.4	0.4	female
2.4	2.3	2.2	2.4	2.3	part-time - total
1.4	1.3	1.2	1.4	1.3	male
1.0	1.0	1.0	1.0	1.0	female
2.9	3.0	3.0	3.2	2.8	Seasonal or casual workers - total
2.2	2.3	2.2	2.4	2.1	male
0.7	0.8	0.8	0.9	0.7	female

LABOUR FORCE on Agricultural Holdings
NORTHERN IRELAND (a)
1982 - 1986 (at June Census)

Table 4.25

	1982	1983	1984	1985	1986
TOTAL LABOUR FORCE	**60.3**	**60.8**	**60.5**	**60.6**	**60.0**
Total farmers,partners,directors (doing farm work)	**36.7**	**36.8**	**36.6**	**36.1**	**35.9**
Whole-time - total	26.3	26.0	25.9	25.3	25.1
principal farmers and partners	na	na	na	na	na
other partners and directors	na	na	na	na	na
Part-time - total	10.4	10.8	10.7	10.7	10.8
principal farmers and partners	na	na	na	na	na
other partners and directors	na	na	na	na	na
Spouses of farmers,partners,directors (b)	**5.2**	**5.0**	**4.9**	**5.3**	**5.4**
Salaried managers	**na**	**na**	**na**	**na**	**na**
Total other workers (c)	**18.4**	**19.0**	**19.1**	**19.3**	**18.7**
Male	15.4	15.9	16.1	16.4	16.0
Female	3.0	3.0	2.9	2.9	2.7
Regular family workers - total	8.5	8.3	8.2	8.1	7.8
whole-time - total	4.6	4.4	4.3	4.1	3.9
male	3.4	3.3	3.3	3.2	3.0
female	1.2	1.1	1.0	0.9	0.9
part-time - total	3.9	3.9	3.9	4.0	3.9
male	3.2	3.2	3.2	3.3	3.3
female	0.7	0.7	0.7	0.7	0.6
Regular hired workers - total	4.0	3.9	3.7	3.6	3.4
whole-time - total	2.5	2.4	2.3	2.2	2.1
male	2.3	2.2	2.1	2.0	2.0
female	0.2	0.2	0.2	0.1	0.1
part-time - total	1.5	1.5	1.4	1.4	1.3
male	1.2	1.3	1.2	1.2	1.1
female	0.2	0.2	0.2	0.2	0.2
Seasonal or casual workers - total	6.0	6.8	7.2	7.7	7.5
male	5.2	5.9	6.3	6.8	6.7
female	0.7	0.8	0.8	0.9	0.8

Notes
(a) Excludes estimates for minor holdings.
(b) Collected as "Wives of farmers, partners and directors (if wives working on the farm)".
(c) Excludes school children and all trainees.
na not available.
Totals may not necessarily agree with the sum of their components due to rounding.

LABOUR FORCE on Agricultural Holdings
NORTHERN IRELAND (a)
1987 - 1991 (at June Census)

thousands

1987	1988	1989	1990	1991	
59.4	**59.5**	**59.1**	**58.4**	**57.2**	**TOTAL LABOUR FORCE**
35.4	**35.7**	**35.0**	**34.6**	**34.3**	**Total farmers,partners,directors (doing farm work)**
24.6	24.6	23.2	22.7	22.3	Whole-time - total
na	*na*	*na*	*na*	*na*	principal farmers and partners
na	*na*	*na*	*na*	*na*	other partners and directors
10.8	11.1	11.8	11.9	12.0	Part-time - total
na	*na*	*na*	*na*	*na*	principal farmers and partners
na	*na*	*na*	*na*	*na*	other partners and directors
5.3	**5.0**	**4.9**	**4.7**	**4.6**	**Spouses of farmers,partners,directors (b)**
na	*na*	*na*	*na*	*na*	**Salaried managers**
18.7	**18.7**	**19.2**	**19.1**	**18.4**	**Total other workers (c)**
16.0	15.7	15.8	15.8	15.1	Male
2.6	3.0	3.3	3.3	3.3	Female
7.5	7.2	7.4	7.2	6.9	Regular family workers - total
3.8	3.5	3.4	3.0	2.9	whole-time - total
2.9	2.5	2.4	2.1	2.0	male
0.8	0.9	1.0	0.9	0.9	female
3.8	3.8	4.1	4.1	4.0	part-time - total
3.2	3.0	3.1	3.2	3.1	male
0.6	0.8	1.0	0.9	1.0	female
3.4	3.4	3.3	3.4	3.3	Regular hired workers - total
2.1	2.1	2.0	2.1	2.0	whole-time - total
1.9	1.9	1.8	1.9	1.9	male
0.1	0.1	0.1	0.1	0.1	female
1.4	1.3	1.4	1.4	1.3	part-time - total
1.2	1.0	1.1	1.1	1.0	male
0.2	0.3	0.3	0.3	0.3	female
7.7	8.1	8.4	8.5	8.1	Seasonal or casual workers - total
6.8	7.2	7.4	7.5	7.1	male
0.8	0.9	1.0	1.0	1.0	female

Chapter 5
Regional and county data; 1991

Contents of regional and county tables

ENGLAND (cont'd)

Regional and County Tables

General Notes

a. Estimates for minor holdings are excluded from all figures. The estimates for Northern Ireland and its counties also exclude estimates for low activity holdings.

b. The England and Wales figures for "Total cereals" exclude maize which is included in "Other crops for stockfeed".

c. "Total crops mainly for stockfeed" in Northern Ireland includes some crops not for stockfeed.

d. Estimates for vegetables exclude potatoes, peas for harvesting dry and mushrooms.

e. "HNS" represents hardy nursery stock.

f. Estimates for "All other grassland" include all grasses 5 years old and over, and sole right rough grazing.

g. "Total labour force" estimates include spouses of farmers, partners and directors doing farm work, and salaried managers.

h. Estimates for partners, directors and salaried managers are included in regular whole-time and part-time workers for Scotland and its regions.

Farm Types and Standard Gross Margins

The classification of farms into broadly homogeneous types is based on a system of estimating the dominant activity on each holding.

The relative importance of different farming activities is measured by Standard Gross Margins (SGMs), which are estimates of the value added by each activity. Conceptually the gross margin is the difference between the value of the output of each activity and the variable costs of the necessary inputs to the activity. A standard value or coefficient for each activity's gross margin is calculated for each major agricultural region. These coefficients are expressed in financial terms: so much for each hectare of wheat or each dairy cow, for instance.

The total SGM for a holding is calculated by multiplying the hectares of each crop grown and the number of each type of livestock by the appropriate SGM coefficient and then adding the components. These elements of the total are examined in relative terms to see if there is a dominant activity, and if so the holding is allocated to it. Note that the allocation of a farm to a type does not mean that its activities are exclusively related to that type.

More detailed analyses of holdings by type are available from the census. Enquiry points are given in Chapter 1.

English Region and County Statistics for Main Holdings
June Census 1991

	ENGLAND			NORTH REGION			Cleveland		
LAND	holdings	hectares	area as % of England	holdings	hectares	area as % of England	holdings	hectares	area as % of England
Total agricultural area	**150 966**	**9 332 046**	*100.0*	**11 605**	**1 042 712**	*11.2*	**433**	**29 716**	*0.3*
Total cereals	**55 369**	**2 929 344**	*100.0*	**3 684**	**146 383**	*5.0*	**271**	**13 083**	*0.4*
Wheat	40 470	1 851 393	*100.0*	1 722	67 268	*3.6*	222	8 511	*0.5*
Barley	41 764	988 283	*100.0*	3 400	72 800	*7.4*	222	4 183	*0.4*
Other cereals	9 649	89 669	*100.0*	662	6 315	*7.0*	49	389	*0.4*
Crops mainly for stockfeed									
Peas for harvesting dry	4 508	68 015	*100.0*	99	1 156	*1.7*	10	88	*0.1*
Field beans	8 332	130 035	*100.0*	99	1 344	*1.0*	20	240	*0.2*
Other crops for stockfeed	14 202	87 330	*100.0*	1 236	5 883	*6.7*	32	66	*0.1*
Other arable crops									
Potatoes	16 187	134 964	*100.0*	677	2 576	*1.9*	48	148	*0.1*
Sugar beet	10 316	195 291	*100.0*	0	0	*0.0*	0	0	*0.0*
Oilseed rape	14 006	387 495	*100.0*	1 015	24 336	*6.3*	137	2 900	*0.7*
Other arable crops	6 249	101 227	*100.0*	144	2 014	*2.0*	10	110	*0.1*
Horticultural crops									
Vegetables	10 000	125 040	*100.0*	154	741	*0.6*	20	51	*0.0*
Orchards and small fruit	6 259	42 172	*100.0*	74	174	*0.4*	*	*	*
HNS, bulbs and flowers	4 417	12 458	*100.0*	100	124	*1.0*	10	5	*0.0*
Glasshouse area	5 565	2 144	*100.0*	118	26	*1.2*	*	*	*
Grassland & all other land									
Grass <5 years old	49 841	849 611	*100.0*	4 648	89 352	*10.5*	215	3 107	*0.4*
All other grassland	126 780	3 761 282	*100.0*	10 934	735 937	*19.6*	374	8 090	*0.2*
All other land	79 269	505 640	*100.0*	5 329	32 667	*6.5*	250	1 822	*0.4*
LIVESTOCK	holdings	number	number as % of England	holdings	number	number as % of England	holdings	number	number as % of England
Total cattle and calves	**77 816**	**6 831 527**	*100.0*	**8 447**	**895 507**	*13.1*	**239**	**19 402**	*0.3*
Dairy cows	27 963	1 936 316	*100.0*	2 962	179 008	*9.2*	74	4 083	*0.2*
Beef cows	34 887	708 203	*100.0*	5 070	153 214	*21.6*	103	2 216	*0.3*
Total pigs	**12 363**	**6 395 099**	*100.0*	**490**	**180 013**	*2.8*	**58**	**30 789**	*0.5*
Breeding herd	8 600	660 693	*100.0*	339	19 832	*3.0*	44	3 193	*0.5*
Total sheep and lambs	**49 162**	**20 249 272**	*100.0*	**7 453**	**5 134 041**	*25.4*	**144**	**51 580**	*0.3*
Breeding flock	46 836	9 107 446	*100.0*	7 278	2 265 975	*24.9*	136	21 591	*0.2*
Total fowls	**24 586**	**95 464 554**	*100.0*	**2 293**	**4 003 054**	*4.2*	**86**	**1 278 610**	*1.3*
Laying birds	22 113	26 692 952	*100.0*	2 165	699 699	*2.6*	79	28 706	*0.1*
Table birds	2 000	55 590 986	*100.0*	91	3 111 963	*5.6*	6	1 249 027	*2.2*
LABOUR	holdings	number	number as % of England	holdings	number	number as % of England	holdings	number	number as % of England
Total labour force	**128 219**	**424 940**	*100.0*	**10 177**	**28 017**	*6.6*	**382**	**1 075**	*0.3*
Farmers, partners & directors	122 573	167 669	*100.0*	9 851	13 898	*8.3*	368	519	*0.3*
Regular whole-time workers	37 900	95 597	*100.0*	2 977	5 047	*5.3*	127	210	*0.2*
Regular part-time workers	22 650	43 589	*100.0*	1 498	2 080	*4.8*	60	80	*0.2*
Seasonal and casual workers	23 221	66 649	*100.0*	1 685	2 876	*4.3*	55	120	*0.2*
ANALYSIS BY TOTAL AREA	holdings	total area hectares	area as % of total	holdings	total area hectares	area as % of total	holdings	total area hectares	area as % of total
Under 20 ha	63 359	496 826	*5.3*	3 108	27 211	*2.6*	149	1 184	*4.0*
20 to <100 ha	61 522	3 045 289	*32.6*	5 424	293 385	*28.1*	196	10 811	*36.4*
100 to <300 ha	21 642	3 493 167	*37.4*	2 519	401 707	*38.5*	76	11 255	*37.9*
300 ha and over	4 443	2 296 764	*24.6*	554	320 409	*30.7*	12	6 466	*21.8*
Total	**150 966**	**9 332 046**	*100.0*	**11 605**	**1 042 712**	*100.0*	**433**	**29 716**	*100.0*
ANALYSIS BY FARM TYPE	holdings	total area hectares	area as % of total	holdings	total area hectares	area as % of total	holdings	total area hectares	area as % of total
Dairying	22 701	1 544 109	*16.5*	2 227	162 479	*15.6*	55	3 022	*10.2*
Cattle & sheep	48 296	2 342 899	*25.1*	6 247	634 348	*60.8*	87	3 782	*12.7*
Cropping	39 264	4 697 213	*50.3*	1 497	208 829	*20.0*	201	20 689	*69.6*
Pigs and poultry	8 534	213 116	*2.3*	316	8 874	*0.9*	26	1 424	*4.8*
Horticulture	9 769	132 005	*1.4*	179	1 708	*0.2*	14	82	*0.3*
Unclassified	22 402	402 704	*4.3*	1 139	26 474	*2.5*	50	718	*2.4*
Total	**150 966**	**9 332 046**	*100.0*	**11 605**	**1 042 712**	*100.0*	**433**	**29 716**	*100.0*

English Region and County Statistics for Main Holdings
June Census 1991

	Cumbria			Durham			Northumberland		
LAND	holdings	hectares	area as % of England	holdings	hectares	area as % of England	holdings	hectares	area as % of England
Total agricultural area	**6 374**	**458 893**	**4.9**	**2 145**	**156 855**	**1.7**	**2 379**	**381 056**	**4.1**
Total cereals	**1 417**	**20 385**	**0.7**	**773**	**33 710**	**1.2**	**1 082**	**72 888**	**2.5**
Wheat	106	1 624	0.1	536	18 832	1.0	758	34 686	1.9
Barley	1 373	18 211	1.8	691	13 901	1.4	996	34 007	3.4
Other cereals	127	551	0.6	159	976	1.1	308	4 195	4.7
Crops mainly for stockfeed									
Peas for harvesting dry	*	*	*	17	169	0.2	*	*	*
Field beans	*	*	*	26	201	0.2	*	*	*
Other crops for stockfeed	646	2 548	2.9	134	515	0.6	401	2 691	3.1
Other arable crops									
Potatoes	346	570	0.4	150	792	0.6	106	984	0.7
Sugar beet	0	0	0.0	0	0	0.0	0	0	0.0
Oilseed rape	13	244	0.1	355	7 575	2.0	437	11 798	3.0
Other arable crops	11	94	0.1	41	794	0.8	70	882	0.9
Horticultural crops									
Vegetables	55	132	0.1	32	139	0.1	36	364	0.3
Orchards and small fruit	*	*	*	14	17	0.0	19	96	0.2
HNS, bulbs and flowers	38	34	0.3	22	35	0.3	16	17	0.1
Glasshouse area	*	*	*	30	7	0.3	18	4	0.2
Grassland & all other land									
Grass <5 years old	2 411	44 888	5.3	792	12 679	1.5	1 130	27 216	3.2
All other grassland	6 112	378 031	10.1	1 973	94 880	2.5	2 249	250 027	6.6
All other land	2 520	11 903	2.4	993	5 341	1.1	1 399	12 323	2.4
LIVESTOCK	holdings	number	number as % of England	holdings	number	number as % of England	holdings	number	number as % of England
Total cattle and calves	**4 921**	**542 854**	**7.9**	**1 466**	**114 729**	**1.7**	**1 688**	**208 427**	**3.1**
Dairy cows	2 337	143 688	7.4	330	16 783	0.9	189	12 430	0.6
Beef cows	2 804	64 549	9.1	915	24 784	3.5	1 179	60 219	8.5
Total pigs	**178**	**67 940**	**1.1**	**144**	**46 197**	**0.7**	**88**	**31 656**	**0.5**
Breeding herd	114	7 298	1.1	110	5 711	0.9	56	3 379	0.5
Total sheep and lambs	**4 335**	**2 698 985**	**13.3**	**1 175**	**698 104**	**3.4**	**1 757**	**1 672 225**	**8.3**
Breeding flock	4 221	1 223 686	13.4	1 148	312 505	3.4	1 734	702 797	7.7
Total fowls	**1 307**	**1 995 211**	**2.1**	**381**	**539 507**	**0.6**	**472**	**119 488**	**0.1**
Laying birds	1 234	396 012	1.5	353	151 800	0.6	455	58 702	0.2
Table birds	48	1 488 874	2.7	25	335 786	0.6	12	38 276	0.1
LABOUR	holdings	number	number as % of England	holdings	number	number as % of England	holdings	number	number as % of England
Total labour force	**5 492**	**14 633**	**3.4**	**1 900**	**5 059**	**1.2**	**2 168**	**6 497**	**1.5**
Farmers, partners & directors	5 357	7 543	4.5	1 857	2 577	1.5	2 043	2 940	1.8
Regular whole-time workers	1 471	2 235	2.3	477	812	0.8	821	1 617	1.7
Regular part-time workers	787	1 057	2.4	254	363	0.8	360	500	1.1
Seasonal and casual workers	878	1 331	2.0	324	618	0.9	385	700	1.1
ANALYSIS BY TOTAL AREA	holdings	total area hectares	area as % of total	holdings	total area hectares	area as % of total	holdings	total area hectares	area as % of total
Under 20 ha	1 735	15 543	3.4	667	5 778	3.7	454	3 951	1.0
20 to <100 ha	3 362	179 805	39.2	1 028	54 777	34.9	720	41 846	11.0
100 to <300 ha	1 122	170 520	37.2	392	62 533	39.9	881	149 891	39.3
300 ha and over	155	93 025	20.3	58	33 767	21.5	324	185 369	48.6
Total	**6 374**	**458 893**	**100.0**	**2 145**	**156 855**	**100.0**	**2 379**	**381 056**	**100.0**
ANALYSIS BY FARM TYPE	holdings	total area hectares	area as % of total	holdings	total area hectares	area as % of total	holdings	total area hectares	area as % of total
Dairying	1 786	133 302	29.0	245	14 699	9.4	124	10 121	2.7
Cattle & sheep	3 644	300 670	65.5	1 060	81 989	52.3	1 395	245 792	64.5
Cropping	92	6 925	1.5	480	52 130	33.2	605	117 938	31.0
Pigs and poultry	138	3 544	0.8	82	2 148	1.4	49	1 418	0.4
Horticulture	71	513	0.1	39	301	0.2	36	375	0.1
Unclassified	643	13 938	3.0	239	5 588	3.6	170	5 413	1.4
Total	**6 374**	**458 893**	**100.0**	**2 145**	**156 855**	**100.0**	**2 379**	**381 056**	**100.0**

English Region and County Statistics for Main Holdings
June Census 1991

	Tyne and Wear			YORKS & HUMBER REGION			Humberside		
LAND	holdings	hectares	area as % of England	holdings	hectares	area as % of England	holdings	hectares	area as % of England
Total agricultural area	274	16 192	0.2	16 853	1 106 276	11.9	3 601	286 937	3.1
Total cereals	141	6 317	0.2	7 455	404 450	13.8	2 410	167 069	5.7
Wheat	100	3 616	0.2	6 024	248 290	13.4	2 191	114 045	6.2
Barley	118	2 497	0.3	6 230	148 670	15.0	1 886	50 428	5.1
Other cereals	19	204	0.2	1 141	7 490	8.4	322	2 595	2.9
Crops mainly for stockfeed									
Peas for harvesting dry	*	*	*	811	9 810	14.4	373	4 251	6.3
Field beans	*	*	*	989	11 352	8.7	407	5 151	4.0
Other crops for stockfeed	23	63	0.1	1 615	5 722	6.6	359	1 289	1.5
Other arable crops									
Potatoes	27	81	0.1	2 572	23 229	17.2	813	8 725	6.5
Sugar beet	0	0	0.0	1 707	24 115	12.3	671	9 534	4.9
Oilseed rape	73	1 819	0.5	2 774	59 543	15.4	1 113	28 047	7.2
Other arable crops	12	134	0.1	706	8 872	8.8	277	3 315	3.3
Horticultural crops									
Vegetables	11	54	0.0	1 019	15 648	12.5	573	11 596	9.3
Orchards and small fruit	4	12	0.0	227	437	1.0	58	96	0.2
HNS, bulbs and flowers	14	34	0.3	267	453	3.6	75	83	0.7
Glasshouse area	14	5	0.2	405	296	13.8	178	192	9.0
Grassland & all other land									
Grass <5 years old	100	1 462	0.2	5 550	70 106	8.3	977	9 876	1.2
All other grassland	226	4 910	0.1	14 002	435 582	11.6	2 425	27 225	0.7
All other land	167	1 278	0.3	8 332	36 661	7.3	2 158	10 487	2.1
LIVESTOCK	holdings	number	number as % of England	holdings	number	number as % of England	holdings	number	number as % of England
Total cattle and calves	133	10 095	0.1	8 933	663 286	9.7	1 234	77 831	1.1
Dairy cows	32	2 024	0.1	2 831	157 242	8.1	207	10 856	0.6
Beef cows	69	1 446	0.2	4 128	80 469	11.4	568	12 086	1.7
Total pigs	22	3 431	0.1	2 403	1 755 462	27.5	771	783 660	12.3
Breeding herd	15	251	0.0	1 794	184 474	27.9	601	84 289	12.8
Total sheep and lambs	42	13 147	0.1	6 488	2 716 706	13.4	738	166 704	0.8
Breeding flock	39	5 396	0.1	6 207	1 196 360	13.1	677	74 974	0.8
Total fowls	47	70 238	0.1	3 287	9 000 526	9.4	527	1 446 622	1.5
Laying birds	44	64 479	0.2	2 988	2 134 738	8.0	460	188 969	0.7
Table birds	0	0	0.0	249	5 628 820	10.1	57	1 021 189	1.8
LABOUR	holdings	number	number as % of England	holdings	number	number as % of England	holdings	number	number as % of England
Total labour force	235	753	0.2	14 636	45 896	10.8	3 193	11 670	2.7
Farmers, partners & directors	226	319	0.2	14 097	19 942	11.9	3 017	4 459	2.7
Regular whole-time workers	81	173	0.2	4 473	10 938	11.4	1 292	3 772	3.9
Regular part-time workers	37	80	0.2	2 157	3 913	9.0	431	729	1.7
Seasonal and casual workers	43	107	0.2	2 442	5 702	8.6	600	1 603	2.4
ANALYSIS BY TOTAL AREA	holdings	total area hectares	area as % of total	holdings	total area hectares	area as % of total	holdings	total area hectares	area as % of total
Under 20 ha	103	755	4.7	6 667	51 447	4.7	1 392	8 408	2.9
20 to <100 ha	118	6 146	38.0	6 907	352 917	31.9	1 246	66 055	23.0
100 to <300 ha	48	7 508	46.4	2 791	448 028	40.5	797	133 131	46.4
300 ha and over	5	1 782	11.0	488	253 884	22.9	166	79 342	27.7
Total	274	16 192	100.0	16 853	1 106 276	100.0	3 601	286 937	100.0
ANALYSIS BY FARM TYPE	holdings	total area hectares	area as % of total	holdings	total area hectares	area as % of total	holdings	total area hectares	area as % of total
Dairying	17	1 335	8.2	1 995	121 263	11.0	115	6 062	2.1
Cattle & sheep	61	2 115	13.1	5 128	291 926	26.4	307	5 199	1.8
Cropping	119	11 148	68.8	5 717	608 379	55.0	2 150	252 980	88.2
Pigs and poultry	21	340	2.1	1 477	46 769	4.2	497	17 433	6.1
Horticulture	19	437	2.7	621	4 951	0.4	230	1 874	0.7
Unclassified	37	818	5.0	1 915	32 988	3.0	302	3 388	1.2
Total	274	16 192	100.0	16 853	1 106 276	100.0	3 601	286 937	100.0

English Region and County Statistics for Main Holdings
June Census 1991

	North Yorkshire			South Yorkshire			West Yorkshire		
LAND	holdings	hectares	area as % of England	holdings	hectares	area as % of England	holdings	hectares	area as % of England
Total agricultural area	**8 513**	**634 913**	**6.8**	**1 473**	**82 684**	**0.9**	**3 266**	**101 742**	**1.1**
Total cereals	**3 815**	**183 890**	**6.3**	**713**	**32 294**	**1.1**	**517**	**21 197**	**0.7**
Wheat	2 925	102 415	5.5	530	19 995	1.1	378	11 835	0.6
Barley	3 362	77 551	7.8	562	11 727	1.2	420	8 964	0.9
Other cereals	654	3 923	4.4	89	573	0.6	76	399	0.4
Crops mainly for stockfeed									
Peas for harvesting dry	318	3 846	5.7	85	1 317	1.9	35	396	0.6
Field beans	378	3 728	2.9	146	1 795	1.4	58	678	0.5
Other crops for stockfeed	1 035	3 732	4.3	109	343	0.4	112	357	0.4
Other arable crops									
Potatoes	1 310	11 617	8.6	241	1 765	1.3	208	1 122	0.8
Sugar beet	917	12 409	6.4	89	1 775	0.9	30	397	0.2
Oilseed rape	1 203	22 555	5.8	270	5 593	1.4	188	3 348	0.9
Other arable crops	289	3 173	3.1	107	2 050	2.0	33	335	0.3
Horticultural crops									
Vegetables	249	1 958	1.6	69	1 141	0.9	128	953	0.8
Orchards and small fruit	105	185	0.4	26	66	0.2	38	91	0.2
HNS, bulbs and flowers	94	234	1.9	37	35	0.3	61	101	0.8
Glasshouse area	143	82	3.8	28	10	0.5	56	13	0.6
Grassland & all other land									
Grass <5 years old	3 303	45 065	5.3	539	6 556	0.8	731	8 609	1.0
All other grassland	7 507	322 537	8.6	1 152	25 273	0.7	2 918	60 546	1.6
All other land	4 314	19 902	3.9	712	2 672	0.5	1 148	3 600	0.7
LIVESTOCK	holdings	number	number as % of England	holdings	number	number as % of England	holdings	number	number as % of England
Total cattle and calves	**5 261**	**436 515**	**6.4**	**734**	**52 226**	**0.8**	**1 704**	**96 714**	**1.4**
Dairy cows	1 904	107 459	5.5	219	12 169	0.6	501	26 758	1.4
Beef cows	2 350	49 586	7.0	329	6 111	0.9	881	12 686	1.8
Total pigs	**1 081**	**761 287**	**11.9**	**181**	**66 082**	**1.0**	**370**	**144 433**	**2.3**
Breeding herd	800	80 895	12.2	141	6 004	0.9	252	13 286	2.0
Total sheep and lambs	**4 522**	**2 205 326**	**10.9**	**287**	**89 967**	**0.4**	**941**	**254 709**	**1.3**
Breeding flock	4 394	967 334	10.6	258	40 299	0.4	878	113 753	1.2
Total fowls	**1 828**	**5 648 292**	**5.9**	**269**	**603 151**	**0.6**	**663**	**1 302 461**	**1.4**
Laying birds	1 672	1 073 912	4.0	240	251 935	0.9	616	619 922	2.3
Table birds	140	3 950 065	7.1	17	313 644	0.6	35	343 922	0.6
LABOUR	holdings	number	number as % of England	holdings	number	number as % of England	holdings	number	number as % of England
Total labour force	**7 612**	**23 356**	**5.5**	**1 232**	**3 667**	**0.9**	**2 599**	**7 203**	**1.7**
Farmers, partners & directors	7 353	10 419	6.2	1 197	1 708	1.0	2 530	3 356	2.0
Regular whole-time workers	2 314	5 121	5.4	329	688	0.7	538	1 357	1.4
Regular part-time workers	1 142	2 128	4.9	186	285	0.7	398	771	1.8
Seasonal and casual workers	1 241	2 781	4.2	219	536	0.8	382	782	1.2
ANALYSIS BY TOTAL AREA	holdings	total area hectares	area as % of total	holdings	total area hectares	area as % of total	holdings	total area hectares	area as % of total
Under 20 ha	2 609	22 168	3.5	655	5 187	6.3	2 011	15 684	15.4
20 to <100 ha	4 004	211 195	33.3	602	29 262	35.4	1 055	46 404	45.6
100 to <300 ha	1 642	258 607	40.7	177	29 508	35.7	175	26 782	26.3
300 ha and over	258	142 943	22.5	39	18 726	22.6	25	12 872	12.7
Total	**8 513**	**634 913**	**100.0**	**1 473**	**82 684**	**100.0**	**3 266**	**101 742**	**100.0**
ANALYSIS BY FARM TYPE	holdings	total area hectares	area as % of total	holdings	total area hectares	area as % of total	holdings	total area hectares	area as % of total
Dairying	1 285	84 680	13.3	175	10 098	12.2	420	20 423	20.1
Cattle & sheep	3 143	235 672	37.1	346	13 425	16.2	1 332	37 631	37.0
Cropping	2 605	271 318	42.7	574	53 378	64.6	388	30 702	30.2
Pigs and poultry	557	22 909	3.6	124	2 211	2.7	299	4 216	4.1
Horticulture	202	1 571	0.2	49	224	0.3	140	1 282	1.3
Unclassified	721	18 763	3.0	205	3 349	4.0	687	7 488	7.4
Total	**8 513**	**634 913**	**100.0**	**1 473**	**82 684**	**100.0**	**3 266**	**101 742**	**100.0**

English Region and County Statistics for Main Holdings
June Census 1991

	EAST MIDLANDS REGION			Derbyshire			Leicestershire		
LAND	holdings	hectares	area as % of England	holdings	hectares	area as % of England	holdings	hectares	area as % of England
Total agricultural area	16 583	1 234 044	13.2	3 767	183 044	2.0	2 881	196 125	2.1
Total cereals	8 745	529 585	18.1	1 041	29 446	1.0	1 371	76 108	2.6
Wheat	7 431	387 721	20.9	573	13 294	0.7	1 128	51 227	2.8
Barley	5 668	131 378	13.3	902	15 216	1.5	995	21 806	2.2
Other cereals	1 200	10 487	11.7	177	936	1.0	342	3 075	3.4
Crops mainly for stockfeed									
Peas for harvesting dry	722	10 124	14.9	21	228	0.3	57	1 014	1.5
Field beans	1 831	32 486	25.0	76	816	0.6	332	5 924	4.6
Other crops for stockfeed	1 075	5 191	5.9	218	756	0.9	176	884	1.0
Other arable crops									
Potatoes	2 849	26 696	19.8	316	1 075	0.8	177	1 239	0.9
Sugar beet	2 541	43 450	22.2	20	337	0.2	47	971	0.5
Oilseed rape	3 280	96 026	24.8	218	4 198	1.1	516	15 837	4.1
Other arable crops	927	15 178	15.0	20	254	0.3	113	1 911	1.9
Horticultural crops									
Vegetables	1 896	37 546	30.0	67	414	0.3	61	167	0.1
Orchards and small fruit	357	777	1.8	34	76	0.2	40	137	0.3
HNS, bulbs and flowers	835	3 244	26.0	51	81	0.7	47	184	1.5
Glasshouse area	727	211	9.9	56	11	0.5	65	16	0.7
Grassland & all other land									
Grass <5 years old	4 712	69 659	8.2	1 176	14 039	1.7	1 083	18 626	2.2
All other grassland	12 944	317 813	8.4	3 497	126 491	3.4	2 517	66 720	1.8
All other land	8 365	46 056	9.1	1 535	4 824	1.0	1 344	6 388	1.3
LIVESTOCK	holdings	number	number as % of England	holdings	number	number as % of England	holdings	number	number as % of England
Total cattle and calves	7 672	636 296	9.3	2 547	202 672	3.0	1 691	160 342	2.3
Dairy cows	2 274	150 629	7.8	1 107	63 302	3.3	554	39 349	2.0
Beef cows	3 369	64 340	9.1	1 107	17 802	2.5	690	11 788	1.7
Total pigs	1 070	621 265	9.7	217	51 345	0.8	182	79 250	1.2
Breeding herd	745	62 655	9.5	136	4 434	0.7	119	6 078	0.9
Total sheep and lambs	4 362	1 724 292	8.5	1 344	461 307	2.3	1 015	450 229	2.2
Breeding flock	4 162	739 104	8.1	1 310	208 073	2.3	967	182 156	2.0
Total fowls	2 295	12 907 516	13.5	777	1 999 111	2.1	377	1 465 190	1.5
Laying birds	2 066	4 386 946	16.4	725	223 991	0.8	343	289 574	1.1
Table birds	152	6 518 928	11.7	38	1 757 260	3.2	25	1 014 567	1.8
LABOUR	holdings	number	number as % of England	holdings	number	number as % of England	holdings	number	number as % of England
Total labour force	14 183	48 792	11.5	3 195	8 619	2.0	2 351	7 079	1.7
Farmers, partners & directors	13 592	19 189	11.4	3 101	4 243	2.5	2 257	3 133	1.9
Regular whole-time workers	4 469	12 431	13.0	771	1 514	1.6	729	1 552	1.6
Regular part-time workers	2 476	5 015	11.5	501	783	1.8	391	770	1.8
Seasonal and casual workers	2 653	6 802	10.2	538	952	1.4	406	803	1.2
ANALYSIS BY TOTAL AREA	holdings	total area hectares	area as % of total	holdings	total area hectares	area as % of total	holdings	total area hectares	area as % of total
Under 20 ha	6 251	50 982	4.1	1 518	14 286	7.8	976	8 564	4.4
20 to <100 ha	6 803	335 764	27.2	1 881	89 431	48.9	1 313	64 867	33.1
100 to <300 ha	2 845	474 760	38.5	332	49 519	27.1	510	82 170	41.9
300 ha and over	684	372 538	30.2	36	29 808	16.3	82	40 523	20.7
Total	16 583	1 234 044	100.0	3 767	183 044	100.0	2 881	196 125	100.0
ANALYSIS BY FARM TYPE	holdings	total area hectares	area as % of total	holdings	total area hectares	area as % of total	holdings	total area hectares	area as % of total
Dairying	1 762	112 180	9.1	975	57 520	31.4	416	27 778	14.2
Cattle & sheep	3 827	155 529	12.6	1 484	72 397	39.6	919	37 135	18.9
Cropping	7 321	894 987	72.5	533	39 050	21.3	906	119 954	61.2
Pigs and poultry	791	17 298	1.4	167	2 975	1.6	106	2 105	1.1
Horticulture	984	17 200	1.4	88	698	0.4	84	551	0.3
Unclassified	1 898	36 850	3.0	520	10 405	5.7	450	8 601	4.4
Total	16 583	1 234 044	100.0	3 767	183 044	100.0	2 881	196 125	100.0

	Lincolnshire			Northamptonshire			Nottinghamshire		
LAND	holdings	hectares	area as % of England	holdings	hectares	area as % of England	holdings	hectares	area as % of England
Total agricultural area	5 916	516 574	5.5	2 053	188 219	2.0	1 966	150 082	1.6
Total cereals	4 030	271 097	9.3	1 043	81 426	2.8	1 260	71 509	2.4
Wheat	3 768	213 669	11.5	922	62 194	3.4	1 040	47 337	2.6
Barley	2 194	54 553	5.5	656	16 918	1.7	921	22 886	2.3
Other cereals	332	2 876	3.2	209	2 313	2.6	140	1 287	1.4
Crops mainly for stockfeed									
Peas for harvesting dry	494	6 507	9.6	35	682	1.0	115	1 693	2.5
Field beans	848	14 261	11.0	297	7 123	5.5	278	4 363	3.4
Other crops for stockfeed	437	1 772	2.0	116	966	1.1	128	812	0.9
Other arable crops									
Potatoes	1 904	18 110	13.4	79	760	0.6	373	5 514	4.1
Sugar beet	2 069	33 535	17.2	27	436	0.2	378	8 171	4.2
Oilseed rape	1 419	41 874	10.8	553	20 099	5.2	574	14 019	3.6
Other arable crops	496	8 148	8.0	146	2 545	2.5	152	2 322	2.3
Horticultural crops									
Vegetables	1 609	34 892	27.9	35	59	0.0	124	2 015	1.6
Orchards and small fruit	187	244	0.6	36	124	0.3	60	198	0.5
HNS, bulbs and flowers	633	2 743	22.0	32	41	0.3	72	194	1.6
Glasshouse area	518	162	7.6	31	8	0.4	57	15	0.7
Grassland & all other land									
Grass <5 years old	1 172	15 169	1.8	657	12 297	1.4	624	9 527	1.1
All other grassland	3 710	48 709	1.3	1 756	51 654	1.4	1 464	24 239	0.6
All other land	3 288	19 351	3.8	1 098	10 000	2.0	1 100	5 493	1.1
LIVESTOCK	holdings	number	number as % of England	holdings	number	number as % of England	holdings	number	number as % of England
Total cattle and calves	1 613	114 103	1.7	997	91 949	1.3	824	67 230	1.0
Dairy cows	229	16 628	0.9	183	14 970	0.8	201	16 380	0.8
Beef cows	808	19 331	2.7	434	9 548	1.3	330	5 871	0.8
Total pigs	361	289 499	4.5	126	62 669	1.0	184	138 502	2.2
Breeding herd	258	31 760	4.8	96	6 471	1.0	136	13 912	2.1
Total sheep and lambs	776	264 563	1.3	899	453 156	2.2	328	95 037	0.5
Breeding flock	720	114 886	1.3	868	193 144	2.1	297	40 845	0.4
Total fowls	564	6 485 651	6.8	284	1 228 234	1.3	293	1 729 330	1.8
Laying birds	477	2 286 927	8.6	259	386 490	1.4	262	1 199 964	4.5
Table birds	52	2 786 702	5.0	15	686 315	1.2	22	274 084	0.5
LABOUR	holdings	number	number as % of England	holdings	number	number as % of England	holdings	number	number as % of England
Total labour force	5 219	21 956	5.2	1 744	5 220	1.2	1 674	5 918	1.4
Farmers, partners & directors	4 987	7 194	4.3	1 637	2 301	1.4	1 610	2 318	1.4
Regular whole-time workers	1 817	6 629	6.9	557	1 176	1.2	595	1 560	1.6
Regular part-time workers	958	2 349	5.4	331	499	1.1	295	614	1.4
Seasonal and casual workers	1 079	3 631	5.4	311	599	0.9	319	817	1.2
ANALYSIS BY TOTAL AREA	holdings	total area hectares	area as % of total	holdings	total area hectares	area as % of total	holdings	total area hectares	area as % of total
Under 20 ha	2 405	17 067	3.3	647	5 507	2.9	705	5 557	3.7
20 to <100 ha	2 054	100 540	19.5	777	40 535	21.5	778	40 392	26.9
100 to <300 ha	1 059	182 607	35.3	528	90 323	48.0	416	70 140	46.7
300 ha and over	398	216 360	41.9	101	51 854	27.5	67	33 992	22.6
Total	5 916	516 574	100.0	2 053	188 219	100.0	1 966	150 082	100.0
ANALYSIS BY FARM TYPE	holdings	total area hectares	area as % of total	holdings	total area hectares	area as % of total	holdings	total area hectares	area as % of total
Dairying	142	7 793	1.5	107	8 565	4.6	122	10 523	7.0
Cattle & sheep	498	9 557	1.9	663	30 023	16.0	263	6 417	4.3
Cropping	3 877	469 884	91.0	886	141 755	75.3	1 119	124 344	82.9
Pigs and poultry	283	6 748	1.3	88	1 461	0.8	147	4 010	2.7
Horticulture	662	14 392	2.8	50	331	0.2	100	1 228	0.8
Unclassified	454	8 200	1.6	259	6 084	3.2	215	3 559	2.4
Total	5 916	516 574	100.0	2 053	188 219	100.0	1 966	150 082	100.0

English Region and County Statistics for Main Holdings
June Census 1991

	EAST ANGLIA REGION			Cambridgeshire			Norfolk		
LAND	holdings	hectares	area as % of England	holdings	hectares	area as % of England	holdings	hectares	area as % of England
Total agricultural area	**12 417**	**1 010 570**	**10.8**	**3 556**	**281 916**	**3.0**	**5 280**	**426 055**	**4.6**
Total cereals	**8 181**	**517 896**	**17.7**	**2 607**	**156 002**	**5.3**	**3 291**	**201 098**	**6.9**
Wheat	7 049	341 927	18.5	2 488	124 411	6.7	2 587	110 500	6.0
Barley	5 604	168 222	17.0	1 333	30 030	3.0	2 553	87 832	8.9
Other cereals	681	7 747	8.6	152	1 561	1.7	289	2 765	3.1
Crops mainly for stockfeed									
Peas for harvesting dry	883	13 104	19.3	387	5 668	8.3	296	3 926	5.8
Field beans	1 959	27 716	21.3	712	10 736	8.3	552	7 346	5.6
Other crops for stockfeed	1 103	6 254	7.2	135	689	0.8	687	3 274	3.7
Other arable crops									
Potatoes	2 633	30 122	22.3	1 111	11 486	8.5	1 127	13 843	10.3
Sugar beet	4 831	103 961	53.2	1 458	23 336	11.9	2 399	56 856	29.1
Oilseed rape	1 675	47 992	12.4	653	20 255	5.2	392	10 263	2.6
Other arable crops	1 269	19 737	19.5	324	4 687	4.6	545	9 059	8.9
Horticultural crops									
Vegetables	1 991	37 921	30.3	505	7 579	6.1	1 007	20 500	16.4
Orchards and small fruit	963	6 121	14.5	306	1 736	4.1	466	2 724	6.5
HNS, bulbs and flowers	616	2 244	18.0	214	762	6.1	261	1 145	9.2
Glasshouse area	483	179	8.3	174	86	4.0	201	61	2.8
Grassland & all other land									
Grass <5 years old	2 212	24 398	2.9	433	3 981	0.5	1 033	12 153	1.4
All other grassland	7 712	104 886	2.8	1 804	22 265	0.6	3 546	53 621	1.4
All other land	7 824	68 042	13.5	1 986	12 650	2.5	3 236	30 186	6.0
LIVESTOCK	holdings	number	number as % of England	holdings	number	number as % of England	holdings	number	number as % of England
Total cattle and calves	**2 965**	**207 953**	**3.0**	**586**	**33 301**	**0.5**	**1 487**	**111 866**	**1.6**
Dairy cows	550	38 811	2.0	64	3 516	0.2	292	21 832	1.1
Beef cows	1 423	31 269	4.4	288	5 738	0.8	738	17 490	2.5
Total pigs	**1 567**	**1 364 343**	**21.3**	**193**	**109 240**	**1.7**	**744**	**617 567**	**9.7**
Breeding herd	1 056	132 951	20.1	148	11 831	1.8	487	62 182	9.4
Total sheep and lambs	**1 274**	**299 778**	**1.5**	**243**	**55 524**	**0.3**	**618**	**171 604**	**0.8**
Breeding flock	1 170	138 097	1.5	215	25 357	0.3	564	77 831	0.9
Total fowls	**1 675**	**14 133 645**	**14.8**	**308**	**2 215 043**	**2.3**	**784**	**6 955 094**	**7.3**
Laying birds	1 395	2 596 053	9.7	261	357 039	1.3	646	1 695 264	6.4
Table birds	256	9 866 965	17.7	41	1 621 364	2.9	110	4 054 052	7.3
LABOUR	holdings	number	number as % of England	holdings	number	number as % of England	holdings	number	number as % of England
Total labour force	**10 825**	**43 059**	**10.1**	**3 145**	**11 200**	**2.6**	**4 550**	**18 925**	**4.5**
Farmers, partners & directors	10 140	14 051	8.4	2 945	4 121	2.5	4 267	5 795	3.5
Regular whole-time workers	3 704	12 700	13.3	982	2 742	2.9	1 562	5 831	6.1
Regular part-time workers	2 067	4 706	10.8	537	1 086	2.5	874	2 072	4.8
Seasonal and casual workers	2 018	7 411	11.1	701	2 072	3.1	757	3 541	5.3
ANALYSIS BY TOTAL AREA	holdings	total area hectares	area as % of total	holdings	total area hectares	area as % of total	holdings	total area hectares	area as % of total
Under 20 ha	5 082	33 871	3.4	1 266	9 235	3.3	2 419	15 355	3.6
20 to <100 ha	4 430	220 513	21.8	1 447	70 875	25.1	1 693	83 226	19.5
100 to <300 ha	2 180	375 159	37.1	643	108 869	38.6	845	147 696	34.7
300 ha and over	725	381 027	37.7	200	92 937	33.0	323	179 778	42.2
Total	**12 417**	**1 010 570**	**100.0**	**3 556**	**281 916**	**100.0**	**5 280**	**426 055**	**100.0**
ANALYSIS BY FARM TYPE	holdings	total area hectares	area as % of total	holdings	total area hectares	area as % of total	holdings	total area hectares	area as % of total
Dairying	220	12 764	1.3	24	1 255	0.4	107	7 032	1.7
Cattle & sheep	1 059	28 032	2.8	197	7 080	2.5	530	14 344	3.4
Cropping	7 830	902 599	89.3	2 589	261 883	92.9	3 138	375 765	88.2
Pigs and poultry	1 127	32 974	3.3	129	2 611	0.9	567	13 970	3.3
Horticulture	1 045	15 678	1.6	357	3 690	1.3	439	7 676	1.8
Unclassified	1 136	18 523	1.8	260	5 397	1.9	499	7 269	1.7
Total	**12 417**	**1 010 570**	**100.0**	**3 556**	**281 916**	**100.0**	**5 280**	**426 055**	**100.0**

	Suffolk			SOUTH EAST REGION			Bedfordshire		
LAND	holdings	hectares	area as % of England	holdings	hectares	area as % of England	holdings	hectares	area as % of England
Total agricultural area	3 581	302 598	3.2	26 202	1 694 182	18.2	1 224	90 141	1.0
Total cereals	2 283	160 795	5.5	8 857	662 806	22.6	697	48 032	1.6
Wheat	1 974	107 016	5.8	7 395	451 566	24.4	601	37 205	2.0
Barley	1 718	50 359	5.1	5 974	188 182	19.0	424	10 049	1.0
Other cereals	240	3 420	3.8	1 721	23 058	25.7	68	778	0.9
Crops mainly for stockfeed									
Peas for harvesting dry	200	3 511	5.2	1 000	20 409	30.0	34	704	1.0
Field beans	695	9 633	7.4	1 851	34 751	26.7	222	5 001	3.8
Other crops for stockfeed	281	2 290	2.6	1 666	16 206	18.6	43	296	0.3
Other arable crops									
Potatoes	395	4 793	3.6	1 620	13 868	10.3	186	716	0.5
Sugar beet	974	23 769	12.2	259	5 058	2.6	26	502	0.3
Oilseed rape	630	17 474	4.5	3 143	106 757	27.6	253	9 318	2.4
Other arable crops	400	5 992	5.9	1 915	35 535	35.1	97	1 776	1.8
Horticultural crops									
Vegetables	479	9 842	7.9	1 881	15 207	12.2	249	1 955	1.6
Orchards and small fruit	191	1 661	3.9	2 197	23 452	55.6	37	96	0.2
HNS, bulbs and flowers	141	337	2.7	1 109	2 639	21.2	47	100	0.8
Glasshouse area	108	32	1.5	1 893	806	37.6	130	47	2.2
Grassland & all other land									
Grass <5 years old	746	8 265	1.0	7 253	143 437	16.9	257	3 311	0.4
All other grassland	2 362	29 000	0.8	21 139	450 239	12.0	828	11 430	0.3
All other land	2 602	25 205	5.0	16 599	163 014	32.2	759	6 857	1.4

	Suffolk			SOUTH EAST REGION			Bedfordshire		
LIVESTOCK	holdings	number	number as % of England	holdings	number	number as % of England	holdings	number	number as % of England
Total cattle and calves	892	62 786	0.9	9 393	781 253	11.4	306	20 978	0.3
Dairy cows	194	13 463	0.7	2 301	200 777	10.4	57	4 687	0.2
Beef cows	397	8 041	1.1	4 230	84 200	11.9	146	2 216	0.3
Total pigs	630	637 536	10.0	1 794	852 672	13.3	67	56 588	0.9
Breeding herd	421	58 938	8.9	1 244	94 615	14.3	49	6 588	1.0
Total sheep and lambs	413	72 650	0.4	6 112	2 172 518	10.7	197	63 776	0.3
Breeding flock	391	34 909	0.4	5 650	973 010	10.7	180	29 030	0.3
Total fowls	583	4 963 508	5.2	3 969	18 406 108	19.3	127	671 135	0.7
Laying birds	488	543 750	2.0	3 553	5 933 147	22.2	115	396 483	1.5
Table birds	105	4 191 549	7.5	313	9 865 840	17.7	7	213 728	0.4

	Suffolk			SOUTH EAST REGION			Bedfordshire		
LABOUR	holdings	number	number as % of England	holdings	number	number as % of England	holdings	number	number as % of England
Total labour force	3 130	12 934	3.0	22 006	89 390	21.0	1 039	3 707	0.9
Farmers, partners & directors	2 928	4 135	2.5	20 430	27 574	16.4	988	1 404	0.8
Regular whole-time workers	1 160	4 127	4.3	7 122	22 339	23.4	327	980	1.0
Regular part-time workers	656	1 548	3.6	4 941	11 866	27.2	191	414	0.9
Seasonal and casual workers	560	1 798	2.7	4 591	18 326	27.5	195	490	0.7

	Suffolk			SOUTH EAST REGION			Bedfordshire		
ANALYSIS BY TOTAL AREA	holdings	total area hectares	area as % of total	holdings	total area hectares	area as % of total	holdings	total area hectares	area as % of total
Under 20 ha	1 397	9 281	3.1	12 733	91 644	5.4	542	3 729	4.1
20 to <100 ha	1 290	66 411	21.9	8 452	407 339	24.0	414	20 188	22.4
100 to <300 ha	692	118 594	39.2	3 920	662 336	39.1	211	36 362	40.3
300 ha and over	202	108 312	35.8	1 097	532 863	31.5	57	29 862	33.1
Total	3 581	302 598	100.0	26 202	1 694 182	100.0	1 224	90 141	100.0

	Suffolk			SOUTH EAST REGION			Bedfordshire		
ANALYSIS BY FARM TYPE	holdings	total area hectares	area as % of total	holdings	total area hectares	area as % of total	holdings	total area hectares	area as % of total
Dairying	89	4 477	1.5	1 566	141 582	8.4	32	2 547	2.8
Cattle & sheep	332	6 608	2.2	6 850	237 972	14.0	151	3 945	4.4
Cropping	2 103	264 950	87.6	7 684	1 125 440	66.4	661	78 752	87.4
Pigs and poultry	431	16 393	5.4	1 585	38 168	2.3	68	876	1.0
Horticulture	249	4 312	1.4	3 334	49 586	2.9	172	1 653	1.8
Unclassified	377	5 857	1.9	5 183	101 434	6.0	140	2 368	2.6
Total	3 581	302 598	100.0	26 202	1 694 182	100.0	1 224	90 141	100.0

English Region and County Statistics for Main Holdings
June Census 1991

	Berkshire			Buckinghamshire			East Sussex		
LAND	holdings	hectares	area as % of England	holdings	hectares	area as % of England	holdings	hectares	area as % of England
Total agricultural area	901	72 552	0.8	1 920	127 418	1.4	2 335	114 111	1.2
Total cereals	304	29 584	1.0	695	42 618	1.5	509	23 472	0.8
Wheat	247	18 638	1.0	586	29 618	1.6	369	14 522	0.8
Barley	228	9 839	1.0	479	11 622	1.2	302	6 849	0.7
Other cereals	57	1 108	1.2	121	1 379	1.5	193	2 101	2.3
Crops mainly for stockfeed									
Peas for harvesting dry	9	142	0.2	11	221	0.3	46	640	0.9
Field beans	37	697	0.5	120	2 673	2.1	76	927	0.7
Other crops for stockfeed	104	1 479	1.7	82	816	0.9	127	1 011	1.2
Other arable crops									
Potatoes	23	112	0.1	*	*	*	86	330	0.2
Sugar beet	0	0	0.0	*	*	*	0	0	0.0
Oilseed rape	78	4 138	1.1	252	7 976	2.1	105	2 583	0.7
Other arable crops	72	1 553	1.5	64	1 008	1.0	111	2 056	2.0
Horticultural crops									
Vegetables	38	333	0.3	44	433	0.3	97	258	0.2
Orchards and small fruit	40	103	0.2	66	222	0.5	175	1 181	2.8
HNS, bulbs and flowers	25	17	0.1	47	54	0.4	74	84	0.7
Glasshouse area	31	10	0.5	57	14	0.7	88	24	1.1
Grassland & all other land									
Grass <5 years old	310	7 850	0.9	594	11 464	1.3	654	12 185	1.4
All other grassland	770	18 000	0.5	1 683	48 426	1.3	2 096	55 159	1.5
All other land	599	8 534	1.7	1 100	11 368	2.2	1 595	14 201	2.8
LIVESTOCK	holdings	number	number as % of England	holdings	number	number as % of England	holdings	number	number as % of England
Total cattle and calves	395	38 818	0.6	984	90 868	1.3	1 031	76 168	1.1
Dairy cows	105	9 911	0.5	229	17 852	0.9	254	19 812	1.0
Beef cows	182	3 959	0.6	427	9 395	1.3	503	7 988	1.1
Total pigs	82	60 888	1.0	115	48 724	0.8	139	28 216	0.4
Breeding herd	63	7 284	1.1	83	5 012	0.8	83	2 589	0.4
Total sheep and lambs	182	71 442	0.4	625	248 696	1.2	840	319 682	1.6
Breeding flock	167	30 901	0.3	593	107 670	1.2	782	144 524	1.6
Total fowls	141	285 730	0.3	288	2 091 668	2.2	392	881 004	0.9
Laying birds	133	163 084	0.6	262	214 583	0.8	355	219 883	0.8
Table birds	7	109 068	0.2	22	1 763 907	3.2	33	554 579	1.0
LABOUR	holdings	number	number as % of England	holdings	number	number as % of England	holdings	number	number as % of England
Total labour force	741	2 516	0.6	1 625	4 755	1.1	1 940	6 920	1.6
Farmers, partners & directors	641	835	0.5	1 515	2 079	1.2	1 840	2 399	1.4
Regular whole-time workers	275	783	0.8	446	980	1.0	470	1 020	1.1
Regular part-time workers	168	283	0.6	343	551	1.3	389	691	1.6
Seasonal and casual workers	132	315	0.5	324	557	0.8	390	2 064	3.1
ANALYSIS BY TOTAL AREA	holdings	total area hectares	area as % of total	holdings	total area hectares	area as % of total	holdings	total area hectares	area as % of total
Under 20 ha	418	3 178	4.4	739	6 525	5.1	1 122	9 397	8.2
20 to <100 ha	282	13 509	18.6	771	38 018	29.8	909	41 962	36.8
100 to <300 ha	146	24 785	34.2	360	58 961	46.3	253	41 538	36.4
300 ha and over	55	31 081	42.8	50	23 914	18.8	51	21 214	18.6
Total	901	72 552	100.0	1 920	127 418	100.0	2 335	114 111	100.0
ANALYSIS BY FARM TYPE	holdings	total area hectares	area as % of total	holdings	total area hectares	area as % of total	holdings	total area hectares	area as % of total
Dairying	74	7 959	11.0	162	13 764	10.8	194	15 420	13.5
Cattle & sheep	245	6 951	9.6	661	30 942	24.3	987	39 244	34.4
Cropping	249	49 256	67.9	511	70 830	55.6	344	44 027	38.6
Pigs and poultry	73	3 300	4.5	107	2 176	1.7	119	1 863	1.6
Horticulture	51	433	0.6	100	1 386	1.1	181	2 685	2.4
Unclassified	209	4 652	6.4	379	8 321	6.5	510	10 871	9.5
Total	901	72 552	100.0	1 920	127 418	100.0	2 335	114 111	100.0

	Essex			Greater London			Hampshire		
LAND	holdings	hectares	area as % of England	holdings	hectares	area as % of England	holdings	hectares	area as % of England
Total agricultural area	**3 718**	**266 366**	**2.9**	**467**	**15 075**	**0.2**	**3 138**	**228 283**	**2.4**
Total cereals	**1 790**	**146 096**	**5.0**	**79**	**3 671**	**0.1**	**877**	**92 320**	**3.2**
Wheat	1 620	111 243	6.0	56	2 357	0.1	705	51 576	2.8
Barley	1 201	33 508	3.4	51	1 053	0.1	708	36 685	3.7
Other cereals	143	1 345	1.5	19	261	0.3	204	4 060	4.5
Crops mainly for stockfeed									
Peas for harvesting dry	276	5 096	7.5	4	82	0.1	179	4 311	6.3
Field beans	455	7 350	5.7	12	187	0.1	103	2 226	1.7
Other crops for stockfeed	156	1 074	1.2	9	80	0.1	324	3 924	4.5
Other arable crops									
Potatoes	405	4 952	3.7	14	101	0.1	*	*	*
Sugar beet	211	4 186	2.1	0	0	0.0	*	*	*
Oilseed rape	815	26 292	6.8	29	647	0.2	226	7 789	2.0
Other arable crops	393	6 166	6.1	13	168	0.2	255	6 017	5.9
Horticultural crops									
Vegetables	315	2 799	2.2	49	487	0.4	237	2 115	1.7
Orchards and small fruit	264	2 426	5.8	30	127	0.3	199	717	1.7
HNS, bulbs and flowers	163	188	1.5	53	63	0.5	130	639	5.1
Glasshouse area	407	183	8.5	82	30	1.4	193	73	3.4
Grassland & all other land									
Grass <5 years old	739	8 973	1.1	87	1 518	0.2	1 006	27 477	3.2
All other grassland	2 563	31 169	0.8	357	6 545	0.2	2 604	55 795	1.5
All other land	2 434	19 418	3.8	234	1 369	0.3	1 918	23 940	4.7
LIVESTOCK	holdings	number	number as % of England	holdings	number	number as % of England	holdings	number	number as % of England
Total cattle and calves	**782**	**52 942**	**0.8**	**121**	**8 487**	**0.1**	**1 335**	**127 374**	**1.9**
Dairy cows	158	11 732	0.6	29	2 129	0.1	376	38 432	2.0
Beef cows	321	5 870	0.8	47	1 048	0.1	597	12 851	1.8
Total pigs	**270**	**128 751**	**2.0**	**45**	**14 984**	**0.2**	**261**	**125 942**	**2.0**
Breeding herd	184	12 559	1.9	28	1 042	0.2	189	15 837	2.4
Total sheep and lambs	**410**	**83 639**	**0.4**	**45**	**6 381**	**0.0**	**594**	**218 662**	**1.1**
Breeding flock	377	38 137	0.4	37	2 871	0.0	544	95 160	1.0
Total fowls	**489**	**3 636 218**	**3.8**	**69**	**61 403**	**0.1**	**482**	**4 751 737**	**5.0**
Laying birds	404	874 972	3.3	61	33 471	0.1	417	1 383 803	5.2
Table birds	68	2 214 152	4.0	*	*	*	46	2 577 162	4.6
LABOUR	holdings	number	number as % of England	holdings	number	number as % of England	holdings	number	number as % of England
Total labour force	**3 161**	**12 380**	**2.9**	**388**	**1 693**	**0.4**	**2 618**	**10 664**	**2.5**
Farmers, partners & directors	2 951	4 199	2.5	354	480	0.3	2 411	3 229	1.9
Regular whole-time workers	1 140	3 106	3.2	126	368	0.4	875	3 236	3.4
Regular part-time workers	744	1 815	4.2	105	254	0.6	583	1 489	3.4
Seasonal and casual workers	640	1 947	2.9	91	407	0.6	503	1 503	2.3
ANALYSIS BY TOTAL AREA	holdings	total area hectares	area as % of total	holdings	total area hectares	area as % of total	holdings	total area hectares	area as % of total
Under 20 ha	1 801	10 091	3.8	295	1 684	11.2	1 654	11 751	5.1
20 to <100 ha	1 020	51 106	19.2	133	6 288	41.7	882	41 593	18.2
100 to <300 ha	718	123 810	46.5	35	5 372	35.6	397	67 626	29.6
300 ha and over	179	81 359	30.5	4	1 732	11.5	205	107 314	47.0
Total	**3 718**	**266 366**	**100.0**	**467**	**15 075**	**100.0**	**3 138**	**228 283**	**100.0**
ANALYSIS BY FARM TYPE	holdings	total area hectares	area as % of total	holdings	total area hectares	area as % of total	holdings	total area hectares	area as % of total
Dairying	81	5 393	2.0	22	1 831	12.1	251	23 317	10.2
Cattle & sheep	451	10 293	3.9	80	2 542	16.9	871	26 771	11.7
Cropping	1 768	233 355	87.6	75	6 932	46.0	795	154 084	67.5
Pigs and poultry	256	4 729	1.8	54	825	5.5	216	6 702	2.9
Horticulture	541	4 127	1.5	120	1 032	6.8	311	3 675	1.6
Unclassified	621	8 469	3.2	116	1 913	12.7	694	13 735	6.0
Total	**3 718**	**266 366**	**100.0**	**467**	**15 075**	**100.0**	**3 138**	**228 283**	**100.0**

	Hertfordshire			Isle of Wight			Kent		
LAND			area as % of			area as % of			area as % of
	holdings	hectares	England	holdings	hectares	England	holdings	hectares	England
Total agricultural area	**1 354**	**105 207**	*1.1*	**514**	**25 808**	*0.3*	**4 322**	**248 489**	*2.7*
Total cereals	**584**	**53 697**	*1.8*	**136**	**6 864**	*0.2*	**1 335**	**79 292**	*2.7*
Wheat	512	36 083	*1.9*	109	5 035	*0.3*	1 087	58 177	*3.1*
Barley	437	16 285	*1.6*	83	1 520	*0.2*	768	17 090	*1.7*
Other cereals	106	1 328	*1.5*	25	309	*0.3*	325	4 025	*4.5*
Crops mainly for stockfeed									
Peas for harvesting dry	35	619	*0.9*	17	412	*0.6*	208	4 041	*5.9*
Field beans	223	4 419	*3.4*	12	205	*0.2*	286	4 968	*3.8*
Other crops for stockfeed	52	467	*0.5*	62	470	*0.5*	219	1 579	*1.8*
Other arable crops									
Potatoes	55	375	*0.3*	38	207	*0.2*	377	4 229	*3.1*
Sugar beet	8	205	*0.1*	0	0	*0.0*	3	12	*0.0*
Oilseed rape	275	9 617	*2.5*	32	1 133	*0.3*	459	15 862	*4.1*
Other arable crops	105	1 871	*1.8*	24	387	*0.4*	370	6 591	*6.5*
Horticultural crops									
Vegetables	48	115	*0.1*	33	306	*0.2*	477	5 145	*4.1*
Orchards and small fruit	44	137	*0.3*	31	60	*0.1*	1 046	17 037	*40.4*
HNS, bulbs and flowers	55	74	*0.6*	13	6	*0.0*	219	496	*4.0*
Glasshouse area	114	57	*2.7*	27	18	*0.8*	224	103	*4.8*
Grassland & all other land									
Grass <5 years old	349	5 739	*0.7*	183	3 117	*0.4*	1 158	18 992	*2.2*
All other grassland	1 076	19 295	*0.5*	435	10 672	*0.3*	3 345	66 383	*1.8*
All other land	841	8 521	*1.7*	305	1 951	*0.4*	2 722	23 758	*4.7*
LIVESTOCK			number as % of			number as % of			number as % of
	holdings	number	England	holdings	number	England	holdings	number	England
Total cattle and calves	**437**	**32 297**	*0.5*	**271**	**21 428**	*0.3*	**1 141**	**76 000**	*1.1*
Dairy cows	93	7 442	*0.4*	102	6 774	*0.3*	217	17 110	*0.9*
Beef cows	200	4 738	*0.7*	107	2 027	*0.3*	568	10 195	*1.4*
Total pigs	**101**	**27 796**	*0.4*	**48**	**6 961**	*0.1*	**239**	**54 129**	*0.8*
Breeding herd	67	3 374	*0.5*	31	862	*0.1*	160	5 587	*0.8*
Total sheep and lambs	**210**	**49 922**	*0.2*	**143**	**30 526**	*0.2*	**1 340**	**589 499**	*2.9*
Breeding flock	194	21 490	*0.2*	136	14 385	*0.2*	1 251	274 334	*3.0*
Total fowls	**215**	**816 389**	*0.9*	**106**	**12 414**	*0.0*	**610**	**2 623 379**	*2.7*
Laying birds	193	115 655	*0.4*	101	11 453	*0.0*	550	1 316 886	*4.9*
Table birds	20	495 366	*0.9*	*	*	*	48	764 210	*1.4*
LABOUR			number as % of			number as % of			number as % of
	holdings	number	England	holdings	number	England	holdings	number	England
Total labour force	**1 130**	**3 768**	*0.9*	**440**	**1 498**	*0.4*	**3 661**	**20 535**	*4.8*
Farmers, partners & directors	1 019	1 419	*0.8*	426	545	*0.3*	3 441	4 648	*2.8*
Regular whole-time workers	412	1 058	*1.1*	126	371	*0.4*	1 162	4 354	*4.6*
Regular part-time workers	249	429	*1.0*	74	164	*0.4*	908	2 867	*6.6*
Seasonal and casual workers	195	408	*0.6*	73	235	*0.4*	909	7 145	*10.7*
ANALYSIS BY TOTAL AREA	holdings	total area hectares	area as % of total	holdings	total area hectares	area as % of total	holdings	total area hectares	area as % of total
Under 20 ha	615	4 018	*3.8*	259	1 979	*7.7*	2 210	16 699	*6.7*
20 to <100 ha	393	19 225	*18.3*	181	8 704	*33.7*	1 431	68 929	*27.7*
100 to <300 ha	267	45 460	*43.2*	62	10 102	*39.1*	534	90 259	*36.3*
300 ha and over	79	36 504	*34.7*	12	5 024	*19.5*	147	72 602	*29.2*
Total	**1 354**	**105 207**	*100.0*	**514**	**25 808**	*100.0*	**4 322**	**248 489**	*100.0*
ANALYSIS BY FARM TYPE	holdings	total area hectares	area as % of total	holdings	total area hectares	area as % of total	holdings	total area hectares	area as % of total
Dairying	54	5 080	*4.8*	90	6 233	*24.1*	132	10 115	*4.1*
Cattle & sheep	249	6 578	*6.3*	173	5 500	*21.3*	1 145	38 438	*15.5*
Cropping	569	86 152	*81.9*	87	11 737	*45.5*	1 126	155 354	*62.5*
Pigs and poultry	81	1 138	*1.1*	22	125	*0.5*	212	3 976	*1.6*
Horticulture	133	798	*0.8*	46	321	*1.2*	937	26 607	*10.7*
Unclassified	268	5 462	*5.2*	96	1 893	*7.3*	770	13 998	*5.6*
Total	**1 354**	**105 207**	*100.0*	**514**	**25 808**	*100.0*	**4 322**	**248 489**	*100.0*

English Region and County Statistics for Main Holdings
June Census 1991

	Oxfordshire			Surrey			West Sussex		
LAND	holdings	hectares	area as % of England	holdings	hectares	area as % of England	holdings	hectares	area as % of England
Total agricultural area	2 110	204 131	2.2	1 874	70 026	0.8	2 325	126 576	1.4
Total cereals	1 052	91 420	3.1	241	10 654	0.4	558	35 086	1.2
Wheat	893	58 102	3.1	153	5 144	0.3	457	23 866	1.3
Barley	813	31 257	3.2	171	3 763	0.4	309	8 662	0.9
Other cereals	181	2 061	2.3	100	1 747	1.9	179	2 558	2.9
Crops mainly for stockfeed									
Peas for harvesting dry	74	1 467	2.2	9	159	0.2	98	2 515	3.7
Field beans	210	4 532	3.5	37	565	0.4	58	1 001	0.8
Other crops for stockfeed	200	1 783	2.0	112	1 159	1.3	176	2 068	2.4
Other arable crops									
Potatoes	110	756	0.6	*	*	*	89	961	0.7
Sugar beet	6	121	0.1	*	*	*	0	0	0.0
Oilseed rape	415	15 734	4.1	61	1 609	0.4	143	4 060	1.0
Other arable crops	215	4 179	4.1	56	824	0.8	140	2 941	2.9
Horticultural crops									
Vegetables	71	328	0.3	80	370	0.3	143	562	0.4
Orchards and small fruit	66	437	1.0	72	376	0.9	127	533	1.3
HNS, bulbs and flowers	50	119	1.0	133	519	4.2	100	279	2.2
Glasshouse area	51	10	0.5	152	54	2.5	337	184	8.6
Grassland & all other land									
Grass <5 years old	840	20 109	2.4	476	8 375	1.0	600	14 326	1.7
All other grassland	1 808	47 367	1.3	1 605	33 969	0.9	1 969	46 029	1.2
All other land	1 379	15 768	3.1	1 204	11 296	2.2	1 509	16 032	3.2

	Oxfordshire			Surrey			West Sussex		
LIVESTOCK	holdings	number	number as % of England	holdings	number	number as % of England	holdings	number	number as % of England
Total cattle and calves	1 052	102 998	1.5	720	50 962	0.7	818	81 933	1.2
Dairy cows	277	26 127	1.3	154	13 703	0.7	250	25 066	1.3
Beef cows	454	10 499	1.5	310	5 643	0.8	368	7 771	1.1
Total pigs	201	218 544	3.4	128	33 467	0.5	98	47 682	0.7
Breeding herd	154	25 751	3.9	85	2 466	0.4	68	5 664	0.9
Total sheep and lambs	610	244 448	1.2	410	91 212	0.5	506	154 633	0.8
Breeding flock	572	105 185	1.2	366	40 565	0.4	451	68 758	0.8
Total fowls	343	861 538	0.9	365	640 842	0.7	342	1 072 651	1.1
Laying birds	315	467 661	1.8	337	185 977	0.7	310	549 236	2.1
Table birds	23	271 376	0.5	17	390 929	0.7	18	511 241	0.9

	Oxfordshire			Surrey			West Sussex		
LABOUR	holdings	number	number as % of England	holdings	number	number as % of England	holdings	number	number as % of England
Total labour force	1 815	6 108	1.4	1 516	5 725	1.3	1 932	9 121	2.1
Farmers, partners & directors	1 677	2 268	1.4	1 381	1 761	1.1	1 786	2 308	1.4
Regular whole-time workers	703	1 798	1.9	435	1 573	1.6	625	2 712	2.8
Regular part-time workers	344	643	1.5	350	912	2.1	493	1 354	3.1
Seasonal and casual workers	363	661	1.0	301	810	1.2	475	1 784	2.7

	Oxfordshire			Surrey			West Sussex		
ANALYSIS BY TOTAL AREA	holdings	total area hectares	area as % of total	holdings	total area hectares	area as % of total	holdings	total area hectares	area as % of total
Under 20 ha	694	5 840	2.9	1 089	7 936	11.3	1 295	8 818	7.0
20 to <100 ha	766	39 631	19.4	621	27 638	39.5	649	30 551	24.1
100 to <300 ha	490	83 980	41.1	139	22 930	32.7	308	51 152	40.4
300 ha and over	160	74 681	36.6	25	11 522	16.5	73	36 055	28.5
Total	2 110	204 131	100.0	1 874	70 026	100.0	2 325	126 576	100.0

	Oxfordshire			Surrey			West Sussex		
ANALYSIS BY FARM TYPE	holdings	total area hectares	area as % of total	holdings	total area hectares	area as % of total	holdings	total area hectares	area as % of total
Dairying	164	15 031	7.4	130	13 495	19.3	180	21 398	16.9
Cattle & sheep	542	23 545	11.5	653	20 464	29.2	642	22 760	18.0
Cropping	861	148 283	72.6	185	19 692	28.1	453	66 984	52.9
Pigs and poultry	122	8 136	4.0	147	2 411	3.4	108	1 911	1.5
Horticulture	104	1 418	0.7	234	2 850	4.1	404	2 603	2.1
Unclassified	317	7 718	3.8	525	11 113	15.9	538	10 921	8.6
Total	2 110	204 131	100.0	1 874	70 026	100.0	2 325	126 576	100.0

English Region and County Statistics for Main Holdings
June Census 1991

	SOUTH WEST REGION			Avon			Cornwall and Isles of Scilly		
LAND	holdings	hectares	area as % of England	holdings	hectares	area as % of England	holdings	hectares	area as % of England
Total agricultural area	35 704	1 822 778	19.5	2 169	85 278	0.9	7 125	278 182	3.0
Total cereals	9 569	354 157	12.1	427	13 625	0.5	2 214	38 301	1.3
Wheat	5 144	183 072	9.9	298	7 470	0.4	455	6 697	0.4
Barley	7 844	152 290	15.4	326	5 742	0.6	1 998	27 964	2.8
Other cereals	2 238	18 796	21.0	51	412	0.5	570	3 639	4.1
Crops mainly for stockfeed									
Peas for harvesting dry	590	8 816	13.0	17	167	0.2	77	895	1.3
Field beans	533	7 740	6.0	25	227	0.2	26	185	0.1
Other crops for stockfeed	5 064	34 769	39.8	204	1 616	1.9	1 008	3 945	4.5
Other arable crops									
Potatoes	2 392	9 134	6.8	117	242	0.2	840	4 048	3.0
Sugar beet	49	875	0.4	0	0	0.0	6	100	0.1
Oilseed rape	771	22 421	5.8	45	897	0.2	30	513	0.1
Other arable crops	748	13 042	12.9	20	286	0.3	80	847	0.8
Horticultural crops									
Vegetables	1 204	4 433	3.5	85	131	0.1	368	1 469	1.2
Orchards and small fruit	1 276	3 922	9.3	86	208	0.5	179	163	0.4
HNS, bulbs and flowers	738	1 876	15.1	52	65	0.5	283	1 185	9.5
Glasshouse area	700	206	9.6	71	17	0.8	179	53	2.5
Grassland &all other land									
Grass <5 years old	14 508	275 361	32.4	661	11 673	1.4	3 256	48 316	5.7
All other grassland	32 573	986 334	26.2	1 950	51 572	1.4	6 440	163 508	4.3
All other land	18 883	99 693	19.7	942	4 552	0.9	3 616	14 655	2.9
LIVESTOCK	holdings	number	number as % of England	holdings	number	number as % of England	holdings	number	number as % of England
Total cattle and calves	22 388	2 116 677	31.0	1 220	111 782	1.6	4 697	394 134	5.8
Dairy cows	9 414	682 394	35.2	545	40 069	2.1	1 781	100 630	5.2
Beef cows	9 603	174 987	24.7	431	5 555	0.8	2 252	38 574	5.4
Total pigs	2 880	866 896	13.6	150	49 725	0.8	652	77 732	1.2
Breeding herd	1 894	91 073	13.8	98	3 467	0.5	418	8 889	1.3
Total sheep and lambs	12 196	4 197 714	20.7	475	105 740	0.5	2 082	656 792	3.2
Breeding flock	11 619	1 999 147	22.0	434	48 273	0.5	1 987	326 697	3.6
Total fowls	5 978	15 976 751	16.7	305	736 479	0.8	1 274	900 977	0.9
Laying birds	5 403	4 707 850	17.6	281	488 420	1.8	1 182	685 752	2.6
Table birds	529	9 027 271	16.2	25	171 525	0.3	83	94 489	0.2
LABOUR	holdings	number	number as % of England	holdings	number	number as % of England	holdings	number	number as % of England
Total labour force	30 052	86 650	20.4	1 706	6 295	1.5	6 000	15 570	3.7
Farmers, partners & directors	28 981	38 165	22.8	1 630	2 190	1.3	5 868	7 598	4.5
Regular whole-time workers	7 775	16 083	16.8	432	993	1.0	1 215	2 222	2.3
Regular part-time workers	5 102	8 091	18.6	318	741	1.7	832	1 234	2.8
Seasonal and casual workers	5 191	11 854	17.8	320	1 724	2.6	915	2 032	3.0
ANALYSIS BY TOTAL AREA	holdings	total area hectares	area as % of total	holdings	total area hectares	area as % of total	holdings	total area hectares	area as % of total
Under 20 ha	14 849	124 061	6.8	1 026	8 003	9.4	3 386	26 919	9.7
20 to <100 ha	16 181	789 517	43.3	937	44 154	51.8	3 092	148 732	53.5
100 to <300 ha	4 111	630 878	34.6	196	28 989	34.0	616	89 460	32.2
300 ha and over	563	278 322	15.3	10	4 131	4.8	31	13 071	4.7
Total	35 704	1 822 778	100.0	2 169	85 278	100.0	7 125	278 182	100.0
ANALYSIS BY FARM TYPE	holdings	total area hectares	area as % of total	holdings	total area hectares	area as % of total	holdings	total area hectares	area as % of total
Dairying	8 303	598 547	32.8	505	34 863	40.9	1 592	95 019	34.2
Cattle & sheep	14 172	581 548	31.9	698	20 806	24.4	3 002	115 325	41.5
Cropping	3 986	479 569	26.3	214	17 651	20.7	687	43 045	15.5
Pigs and poultry	1 563	37 248	2.0	111	2 136	2.5	272	4 095	1.5
Horticulture	1 488	17 287	0.9	127	778	0.9	391	5 660	2.0
Unclassified	6 192	108 579	6.0	514	9 045	10.6	1 181	15 039	5.4
Total	35 704	1 822 778	100.0	2 169	85 278	100.0	7 125	278 182	100.0

	Devon			Dorset			Gloucestershire		
LAND	holdings	hectares	area as % of England	holdings	hectares	area as % of England	holdings	hectares	area as % of England
Total agricultural area	**10 866**	**514 383**	**5.5**	**3 011**	**197 035**	**2.1**	**3 348**	**204 711**	**2.2**
Total cereals	**2 947**	**58 549**	**2.0**	**674**	**47 604**	**1.6**	**1 007**	**61 811**	**2.1**
Wheat	1 157	17 376	0.9	558	28 909	1.6	806	37 048	2.0
Barley	2 594	35 817	3.6	488	16 410	1.7	746	22 887	2.3
Other cereals	921	5 355	6.0	126	2 286	2.5	169	1 876	2.1
Crops mainly for stockfeed									
Peas for harvesting dry	172	1 743	2.6	88	1 911	2.8	44	686	1.0
Field beans	62	598	0.5	29	527	0.4	133	2 536	2.0
Other crops for stockfeed	1 703	8 439	9.7	517	6 243	7.1	378	3 445	3.9
Other arable crops									
Potatoes	724	1 520	1.1	*	*	*	182	835	0.6
Sugar beet	5	64	0.0	*	*	*	10	154	0.1
Oilseed rape	52	1 175	0.3	42	1 551	0.4	262	8 499	2.2
Other arable crops	99	979	1.0	73	1 509	1.5	206	4 868	4.8
Horticultural crops									
Vegetables	267	1 049	0.8	102	188	0.2	140	524	0.4
Orchards and small fruit	325	643	1.5	91	226	0.5	177	935	2.2
HNS, bulbs and flowers	139	213	1.7	69	102	0.8	58	93	0.7
Glasshouse area	124	39	1.8	67	14	0.7	121	38	1.8
Grassland &all other land									
Grass <5 years old	4 698	72 172	8.5	1 295	38 901	4.6	1 218	24 889	2.9
All other grassland	10 187	339 513	9.0	2 739	87 035	2.3	2 986	79 258	2.1
All other land	6 433	27 688	5.5	1 701	10 901	2.2	1 814	16 142	3.2
LIVESTOCK	holdings	number	number as % of England	holdings	number	number as % of England	holdings	number	number as % of England
Total cattle and calves	**7 328**	**649 383**	**9.5**	**1 977**	**232 188**	**3.4**	**1 752**	**158 713**	**2.3**
Dairy cows	2 921	181 522	9.4	1 009	98 177	5.1	627	47 840	2.5
Beef cows	3 482	63 239	8.9	672	11 712	1.7	772	13 408	1.9
Total pigs	**923**	**182 797**	**2.9**	**247**	**177 910**	**2.8**	**213**	**52 723**	**0.8**
Breeding herd	587	19 970	3.0	170	17 319	2.6	147	6 748	1.0
Total sheep and lambs	**5 230**	**1 926 886**	**9.5**	**768**	**226 672**	**1.1**	**1 130**	**421 946**	**2.1**
Breeding flock	5 045	921 328	10.1	724	105 378	1.2	1 085	189 406	2.1
Total fowls	**2 063**	**3 911 007**	**4.1**	**456**	**1 691 368**	**1.8**	**480**	**1 334 300**	**1.4**
Laying birds	1 893	1 247 897	4.7	410	456 225	1.7	427	452 697	1.7
Table birds	174	2 070 086	3.7	41	892 766	1.6	38	546 763	1.0
LABOUR	holdings	number	number as % of England	holdings	number	number as % of England	holdings	number	number as % of England
Total labour force	**9 478**	**25 008**	**5.9**	**2 614**	**8 444**	**2.0**	**2 748**	**8 309**	**2.0**
Farmers, partners & directors	9 233	11 981	7.1	2 473	3 336	2.0	2 624	3 474	2.1
Regular whole-time workers	2 052	3 680	3.8	857	2 179	2.3	811	1 753	1.8
Regular part-time workers	1 572	2 241	5.1	562	906	2.1	499	902	2.1
Seasonal and casual workers	1 708	3 003	4.5	447	867	1.3	480	1 096	1.6
ANALYSIS BY TOTAL AREA	holdings	total area hectares	area as % of total	holdings	total area hectares	area as % of total	holdings	total area hectares	area as % of total
Under 20 ha	3 944	37 599	7.3	1 209	9 190	4.7	1 441	11 496	5.6
20 to <100 ha	5 719	280 530	54.5	1 234	60 640	30.8	1 344	66 009	32.2
100 to <300 ha	1 139	164 039	31.9	459	73 477	37.3	460	74 450	36.4
300 ha and over	64	32 215	6.3	109	53 729	27.3	103	52 758	25.8
Total	**10 866**	**514 383**	**100.0**	**3 011**	**197 035**	**100.0**	**3 348**	**204 711**	**100.0**
ANALYSIS BY FARM TYPE	holdings	total area hectares	area as % of total	holdings	total area hectares	area as % of total	holdings	total area hectares	area as % of total
Dairying	2 521	161 645	31.4	890	87 849	44.6	531	42 724	20.9
Cattle & sheep	5 348	251 302	48.9	1 037	36 203	18.4	1 169	42 725	20.9
Cropping	737	58 724	11.4	329	56 985	28.9	657	101 870	49.8
Pigs and poultry	441	10 729	2.1	157	5 481	2.8	142	2 686	1.3
Horticulture	283	3 027	0.6	140	1 366	0.7	204	2 286	1.1
Unclassified	1 536	28 957	5.6	458	9 150	4.6	645	12 421	6.1
Total	**10 866**	**514 383**	**100.0**	**3 011**	**197 035**	**100.0**	**3 348**	**204 711**	**100.0**

English Region and County Statistics for Main Holdings
June Census 1991

LAND	Somerset			Wiltshire			WEST MIDLANDS REGION		
	holdings	hectares	area as % of England	holdings	hectares	area as % of England	holdings	hectares	area as % of England
Total agricultural area	**6 049**	**275 295**	**2.9**	**3 136**	**267 894**	**2.9**	**19 403**	**970 904**	**10.4**
Total cereals	**1 244**	**38 278**	**1.3**	**1 056**	**95 991**	**3.3**	**6 657**	**260 336**	**8.9**
Wheat	960	24 977	1.3	910	60 594	3.3	4 538	148 182	8.0
Barley	863	11 681	1.2	829	31 789	3.2	5 182	98 210	9.9
Other cereals	212	1 619	1.8	189	3 609	4.0	1 734	13 944	15.6
Crops mainly for stockfeed									
Peas for harvesting dry	73	736	1.1	119	2 679	3.9	298	3 725	5.5
Field beans	168	1 859	1.4	90	1 809	1.4	836	12 049	9.3
Other crops for stockfeed	866	6 935	7.9	388	4 146	4.7	1 847	9 913	11.4
Other arable crops									
Potatoes	341	1 612	1.2	*	*	*	2 045	20 911	15.5
Sugar beet	23	391	0.2	*	*	*	851	17 107	8.8
Oilseed rape	77	1 403	0.4	263	8 383	2.2	1 071	25 915	6.7
Other arable crops	79	795	0.8	191	3 758	3.7	496	6 357	6.3
Horticultural crops									
Vegetables	161	713	0.6	81	360	0.3	917	6 117	4.9
Orchards and small fruit	348	1 630	3.9	70	117	0.3	1 049	6 975	16.5
HNS, bulbs and flowers	95	131	1.1	42	87	0.7	381	1 274	10.2
Glasshouse area	90	31	1.4	48	14	0.7	487	161	7.5
Grassland &all other land									
Grass <5 years old	2 137	41 558	4.9	1 243	37 852	4.5	7 336	123 548	14.5
All other grassland	5 460	168 655	4.5	2 811	96 792	2.6	17 089	431 768	11.5
All other land	2 628	10 567	2.1	1 749	15 187	3.0	9 381	44 748	8.8

LIVESTOCK	Somerset			Wiltshire			WEST MIDLANDS REGION		
	holdings	number	number as % of England	holdings	number	number as % of England	holdings	number	number as % of England
Total cattle and calves	**3 630**	**357 557**	**5.2**	**1 784**	**212 920**	**3.1**	**11 218**	**928 223**	**13.6**
Dairy cows	1 685	136 475	7.0	846	77 681	4.0	4 040	274 756	14.2
Beef cows	1 369	25 467	3.6	625	17 032	2.4	4 936	86 371	12.2
Total pigs	**463**	**190 136**	**3.0**	**232**	**135 873**	**2.1**	**1 259**	**425 510**	**6.7**
Breeding herd	313	18 058	2.7	161	16 622	2.5	897	44 341	6.7
Total sheep and lambs	**1 850**	**601 975**	**3.0**	**661**	**257 703**	**1.3**	**7 911**	**2 886 097**	**14.3**
Breeding flock	1 742	291 863	3.2	602	116 202	1.3	7 599	1 293 567	14.2
Total fowls	**927**	**3 516 769**	**3.7**	**473**	**3 885 851**	**4.1**	**3 135**	**13 614 722**	**14.3**
Laying birds	802	409 954	1.5	408	966 905	3.6	2 825	3 803 981	14.3
Table birds	104	2 670 988	4.8	64	2 580 654	4.6	253	8 171 116	14.7

LABOUR	Somerset			Wiltshire			WEST MIDLANDS REGION		
	holdings	number	number as % of England	holdings	number	number as % of England	holdings	number	number as % of England
Total labour force	**4 885**	**14 605**	**3.4**	**2 621**	**8 419**	**2.0**	**16 339**	**50 891**	**12.0**
Farmers, partners & directors	4 724	6 327	3.8	2 429	3 259	1.9	15 726	21 267	12.7
Regular whole-time workers	1 362	2 778	2.9	1 046	2 478	2.6	4 510	9 455	9.9
Regular part-time workers	824	1 259	2.9	495	808	1.9	2 685	4 558	10.5
Seasonal and casual workers	883	2 309	3.5	438	823	1.2	2 890	9 198	13.8

ANALYSIS BY TOTAL AREA	Somerset			Wiltshire			WEST MIDLANDS REGION		
	holdings	total area hectares	area as % of total	holdings	total area hectares	area as % of total	holdings	total area hectares	area as % of total
Under 20 ha	2 654	21 318	7.7	1 189	9 536	3.6	8 565	70 333	7.2
20 to <100 ha	2 679	130 754	47.5	1 176	58 700	21.9	8 050	398 646	41.1
100 to <300 ha	669	102 004	37.1	572	98 460	36.8	2 531	390 846	40.3
300 ha and over	47	21 220	7.7	199	101 198	37.8	257	111 078	11.4
Total	**6 049**	**275 295**	**100.0**	**3 136**	**267 894**	**100.0**	**19 403**	**970 904**	**100.0**

ANALYSIS BY FARM TYPE	Somerset			Wiltshire			WEST MIDLANDS REGION		
	holdings	total area hectares	area as % of total	holdings	total area hectares	area as % of total	holdings	total area hectares	area as % of total
Dairying	1 554	111 288	40.4	710	65 158	24.3	3 421	214 512	22.1
Cattle & sheep	2 142	87 334	31.7	776	27 854	10.4	7 366	270 294	27.8
Cropping	605	46 348	16.8	757	154 946	57.8	3 833	401 620	41.4
Pigs and poultry	270	4 718	1.7	170	7 405	2.8	882	20 785	2.1
Horticulture	247	3 302	1.2	96	869	0.3	1 085	16 129	1.7
Unclassified	1 231	22 306	8.1	627	11 662	4.4	2 816	47 564	4.9
Total	**6 049**	**275 295**	**100.0**	**3 136**	**267 894**	**100.0**	**19 403**	**970 904**	**100.0**

	Hereford and Worcester			Shropshire			Staffordshire		
LAND	holdings	hectares	area as % of England	holdings	hectares	area as % of England	holdings	hectares	area as % of England
Total agricultural area	**6 793**	**313 629**	**3.4**	**4 933**	**288 062**	**3.1**	**4 741**	**198 287**	**2.1**
Total cereals	**2 396**	**79 835**	**2.7**	**1 951**	**77 329**	**2.6**	**1 054**	**40 174**	**1.4**
Wheat	1 715	47 773	2.6	1 147	36 573	2.0	673	20 736	1.1
Barley	1 742	25 944	2.6	1 701	36 282	3.7	871	18 344	1.9
Other cereals	800	6 118	6.8	531	4 474	5.0	160	1 094	1.2
Crops mainly for stockfeed									
Peas for harvesting dry	129	1 562	2.3	80	1 006	1.5	35	396	0.6
Field beans	354	4 746	3.6	133	1 640	1.3	86	1 236	1.0
Other crops for stockfeed	751	3 886	4.4	612	3 294	3.8	308	1 813	2.1
Other arable crops									
Potatoes	839	7 389	5.5	540	7 157	5.3	348	3 734	2.8
Sugar beet	294	4 492	2.3	432	9 963	5.1	110	2 331	1.2
Oilseed rape	322	6 445	1.7	185	3 780	1.0	166	3 849	1.0
Other arable crops	261	2 941	2.9	83	978	1.0	42	501	0.5
Horticultural crops									
Vegetables	576	3 121	2.5	110	832	0.7	107	836	0.7
Orchards and small fruit	827	6 029	14.3	71	164	0.4	65	394	0.9
HNS, bulbs and flowers	151	662	5.3	64	173	1.4	81	235	1.9
Glasshouse area	264	104	4.9	54	14	0.7	80	17	0.8
Grassland &all other land									
Grass <5 years old	2 522	38 972	4.6	2 281	41 520	4.9	1 481	24 699	2.9
All other grassland	5 835	136 172	3.6	4 460	129 364	3.4	4 316	109 946	2.9
All other land	3 549	17 273	3.4	2 442	10 848	2.1	1 915	8 126	1.6
LIVESTOCK	holdings	number	number as % of England	holdings	number	number as % of England	holdings	number	number as % of England
Total cattle and calves	**3 396**	**239 989**	**3.5**	**3 194**	**304 435**	**4.5**	**3 186**	**269 812**	**3.9**
Dairy cows	741	50 107	2.6	1 312	95 995	5.0	1 564	100 253	5.2
Beef cows	1 672	28 473	4.0	1 400	29 011	4.1	1 215	18 554	2.6
Total pigs	**370**	**108 712**	**1.7**	**350**	**133 449**	**2.1**	**350**	**120 279**	**1.9**
Breeding herd	244	10 819	1.6	257	15 243	2.3	251	11 900	1.8
Total sheep and lambs	**3 287**	**1 240 272**	**6.1**	**2 280**	**948 020**	**4.7**	**1 301**	**315 469**	**1.6**
Breeding flock	3 179	561 121	6.2	2 202	432 021	4.7	1 219	137 952	1.5
Total fowls	**1 078**	**6 347 069**	**6.6**	**910**	**4 183 693**	**4.4**	**708**	**1 366 728**	**1.4**
Laying birds	954	776 566	2.9	830	1 876 694	7.0	644	683 927	2.6
Table birds	102	4 935 620	8.9	69	1 758 698	3.2	41	549 589	1.0
LABOUR	holdings	number	number as % of England	holdings	number	number as % of England	holdings	number	number as % of England
Total labour force	**5 800**	**18 978**	**4.5**	**4 218**	**12 713**	**3.0**	**3 854**	**11 286**	**2.7**
Farmers, partners & directors	5 563	7 479	4.5	4 080	5 574	3.3	3 734	5 016	3.0
Regular whole-time workers	1 471	3 158	3.3	1 320	2 674	2.8	975	1 906	2.0
Regular part-time workers	948	1 723	4.0	723	1 165	2.7	589	947	2.2
Seasonal and casual workers	1 098	4 327	6.5	702	1 562	2.3	615	1 942	2.9
ANALYSIS BY TOTAL AREA	holdings	total area hectares	area as % of total	holdings	total area hectares	area as % of total	holdings	total area hectares	area as % of total
Under 20 ha	3 201	24 911	7.9	1 926	15 993	5.6	2 210	19 498	9.8
20 to <100 ha	2 681	133 434	42.5	2 109	109 744	38.1	2 072	95 670	48.2
100 to <300 ha	847	127 523	40.7	815	126 601	43.9	*	*	*
300 ha and over	64	27 761	8.9	83	35 723	12.4	*	*	*
Total	**6 793**	**313 629**	**100.0**	**4 933**	**288 062**	**100.0**	**4 741**	**198 287**	**100.0**
ANALYSIS BY FARM TYPE	holdings	total area hectares	area as % of total	holdings	total area hectares	area as % of total	holdings	total area hectares	area as % of total
Dairying	569	38 106	12.1	1 100	73 267	25.4	1 436	80 109	40.4
Cattle & sheep	2 848	107 402	34.2	1 992	87 981	30.5	1 627	44 080	22.2
Cropping	1 466	133 850	42.7	902	109 298	37.9	538	55 157	27.8
Pigs and poultry	274	6 933	2.2	226	5 773	2.0	205	4 729	2.4
Horticulture	697	11 694	3.7	93	846	0.3	141	1 367	0.7
Unclassified	939	15 645	5.0	620	10 897	3.8	794	12 845	6.5
Total	**6 793**	**313 629**	**100.0**	**4 933**	**288 062**	**100.0**	**4 741**	**198 287**	**100.0**

English Region and County Statistics for Main Holdings
June Census 1991

	Warwickshire			West Midlands			NORTH WEST REGION		
LAND	holdings	hectares	area as % of England	holdings	hectares	area as % of England	holdings	hectares	area as % of England
Total agricultural area	**2 445**	**154 964**	*1.7*	**491**	**15 962**	*0.2*	**12 199**	**450 581**	*4.8*
Total cereals	**1 101**	**58 217**	*2.0*	**155**	**4 781**	*0.2*	**2 221**	**53 731**	*1.8*
Wheat	918	40 833	2.2	85	2 267	0.1	1 167	23 369	1.3
Barley	735	15 390	1.6	133	2 251	0.2	1 862	28 531	2.9
Other cereals	206	1 995	2.2	37	264	0.3	272	1 831	2.0
Crops mainly for stockfeed									
Peas for harvesting dry	48	701	1.0	6	60	0.1	105	872	1.3
Field beans	253	4 291	3.3	10	136	0.1	234	2 597	2.0
Other crops for stockfeed	140	801	0.9	36	118	0.1	596	3 394	3.9
Other arable crops									
Potatoes	253	2 288	1.7	65	342	0.3	1 399	8 429	6.2
Sugar beet	7	150	0.1	8	171	0.1	78	726	0.4
Oilseed rape	374	11 306	2.9	24	535	0.1	277	4 504	1.2
Other arable crops	103	1 868	1.8	7	69	0.1	44	490	0.5
Horticultural crops									
Vegetables	99	1 114	0.9	25	214	0.2	938	7 428	5.9
Orchards and small fruit	69	364	0.9	17	25	0.1	116	313	0.7
HNS, bulbs and flowers	58	142	1.1	27	61	0.5	371	606	4.9
Glasshouse area	61	20	0.9	28	6	0.3	752	259	12.1
Grassland &all other land									
Grass <5 years old	890	16 117	1.9	162	2 241	0.3	3 622	53 750	6.3
All other grassland	2 068	49 788	1.3	410	6 499	0.2	10 387	298 724	7.9
All other land	1 256	7 795	1.5	219	706	0.1	4 916	14 758	2.9

	holdings	number	number as % of England	holdings	number	number as % of England	holdings	number	number as % of England
LIVESTOCK									
Total cattle and calves	**1 231**	**99 675**	*1.5*	**211**	**14 312**	*0.2*	**6 800**	**602 332**	*8.8*
Dairy cows	364	24 982	1.3	59	3 419	0.2	3 591	252 699	13.1
Beef cows	547	8 989	1.3	102	1 344	0.2	2 128	33 353	4.7
Total pigs	**145**	**48 117**	*0.8*	**44**	**14 953**	*0.2*	**900**	**328 938**	*5.1*
Breeding herd	109	4 895	0.7	36	1 484	0.2	631	30 752	4.7
Total sheep and lambs	**938**	**356 990**	*1.8*	**105**	**25 346**	*0.1*	**3 366**	**1 118 126**	*5.5*
Breeding flock	904	151 442	1.7	95	11 031	0.1	3 151	502 186	5.5
Total fowls	**360**	**1 528 841**	*1.6*	**79**	**188 391**	*0.2*	**1 954**	**7 422 232**	*7.8*
Laying birds	325	357 932	1.3	72	108 862	0.4	1 718	2 430 538	9.1
Table birds	34	903 101	1.6	7	24 108	0.0	157	3 400 083	6.1

	holdings	number	number as % of England	holdings	number	number as % of England	holdings	number	number as % of England
LABOUR									
Total labour force	**2 076**	**6 492**	*1.5*	**391**	**1 422**	*0.3*	**10 001**	**32 245**	*7.6*
Farmers, partners & directors	1 972	2 678	1.6	377	520	0.3	9 756	13 583	8.1
Regular whole-time workers	642	1 476	1.5	102	241	0.3	2 870	6 604	6.9
Regular part-time workers	356	592	1.4	69	131	0.3	1 724	3 360	7.7
Seasonal and casual workers	396	981	1.5	79	386	0.6	1 751	4 480	6.7

	holdings	total area hectares	area as % of total	holdings	total area hectares	area as % of total	holdings	total area hectares	area as % of total
ANALYSIS BY TOTAL AREA									
Under 20 ha	943	7 827	5.1	285	2 105	13.2	6 104	47 276	10.5
20 to <100 ha	1 014	52 027	33.6	174	7 771	48.7	5 275	247 209	54.9
100 to <300 ha	432	69 661	45.0	*	*	*	745	109 453	24.3
300 ha and over	56	25 450	16.4	*	*	*	75	46 644	10.4
Total	**2 445**	**154 964**	*100.0*	**491**	**15 962**	*100.0*	**12 199**	**450 581**	*100.0*

	holdings	total area hectares	area as % of total	holdings	total area hectares	area as % of total	holdings	total area hectares	area as % of total
ANALYSIS BY FARM TYPE									
Dairying	268	19 578	12.6	48	3 453	21.6	3 207	180 781	40.1
Cattle & sheep	754	27 171	17.5	145	3 661	22.9	3 647	143 251	31.8
Cropping	829	97 162	62.7	98	6 154	38.6	1 396	75 791	16.8
Pigs and poultry	130	2 650	1.7	47	699	4.4	793	11 001	2.4
Horticulture	115	1 869	1.2	39	353	2.2	1 033	9 466	2.1
Unclassified	349	6 535	4.2	114	1 643	10.3	2 123	30 292	6.7
Total	**2 445**	**154 964**	*100.0*	**491**	**15 962**	*100.0*	**12 199**	**450 581**	*100.0*

	Cheshire			Greater Manchester			Lancashire		
LAND			area as % of			area as % of			area as % of
	holdings	hectares	England	holdings	hectares	England	holdings	hectares	England
Total agricultural area	**4 397**	**167 905**	*1.8*	**1 556**	**41 898**	*0.4*	**5 733**	**220 831**	*2.4*
Total cereals	**1 127**	**23 752**	*0.8*	**185**	**5 516**	*0.2*	**670**	**16 265**	*0.6*
Wheat	529	10 016	0.5	102	1 900	0.1	368	7 840	0.4
Barley	938	13 005	1.3	167	3 444	0.3	562	7 846	0.8
Other cereals	131	731	0.8	23	172	0.2	80	579	0.6
Crops mainly for stockfeed									
Peas for harvesting dry	16	122	0.2	5	45	0.1	43	303	0.4
Field beans	43	520	0.4	25	250	0.2	115	1 285	1.0
Other crops for stockfeed	371	2 615	3.0	35	67	0.1	144	483	0.6
Other arable crops									
Potatoes	616	4 193	3.1	110	585	0.4	531	2 718	2.0
Sugar beet	17	147	0.1	7	47	0.0	30	309	0.2
Oilseed rape	114	2 072	0.5	41	572	0.1	77	1 240	0.3
Other arable crops	15	262	0.3	6	32	0.0	14	131	0.1
Horticultural crops									
Vegetables	138	633	0.5	73	730	0.6	587	5 024	4.0
Orchards and small fruit	56	190	0.5	8	16	0.0	37	48	0.1
HNS, bulbs and flowers	139	307	2.5	55	75	0.6	140	164	1.3
Glasshouse area	143	41	1.9	67	20	0.9	488	185	8.6
Grassland &all other land									
Grass <5 years old	1 923	32 191	3.8	344	3 543	0.4	1 207	16 024	1.9
All other grassland	3 837	94 626	2.5	1 369	28 885	0.8	4 854	170 981	4.5
All other land	1 951	6 236	1.2	554	1 516	0.3	2 128	5 672	1.1
LIVESTOCK			number as % of			number as % of			number as % of
	holdings	number	England	holdings	number	England	holdings	number	England
Total cattle and calves	**2 717**	**274 671**	*4.0*	**794**	**40 469**	*0.6*	**3 144**	**275 023**	*4.0*
Dairy cows	1 631	137 139	7.1	262	12 674	0.7	1 659	99 889	5.2
Beef cows	691	11 194	1.6	355	4 491	0.6	1 036	16 649	2.4
Total pigs	**280**	**105 997**	*1.7*	**146**	**28 968**	*0.5*	**439**	**185 107**	*2.9*
Breeding herd	206	10 343	1.6	98	1 715	0.3	304	17 888	2.7
Total sheep and lambs	**947**	**229 781**	*1.1*	**305**	**77 251**	*0.4*	**2 079**	**802 955**	*4.0*
Breeding flock	866	105 516	1.2	269	37 075	0.4	1 983	356 137	3.9
Total fowls	**616**	**2 435 331**	*2.6*	**308**	**550 892**	*0.6*	**969**	**3 988 747**	*4.2*
Laying birds	524	663 548	2.5	281	257 017	1.0	860	1 273 554	4.8
Table birds	58	1 232 193	2.2	14	198 141	0.4	80	1 855 844	3.3
LABOUR			number as % of			number as % of			number as % of
	holdings	number	England	holdings	number	England	holdings	number	England
Total labour force	**3 711**	**11 715**	*2.8*	**1 217**	**3 653**	*0.9*	**4 647**	**15 307**	*3.6*
Farmers, partners & directors	3 604	4 895	2.9	1 194	1 666	1.0	4 548	6 445	3.8
Regular whole-time workers	1 225	2 648	2.8	280	606	0.6	1 213	2 971	3.1
Regular part-time workers	635	1 041	2.4	203	334	0.8	807	1 815	4.2
Seasonal and casual workers	625	1 612	2.4	216	587	0.9	820	2 009	3.0
ANALYSIS BY TOTAL AREA	holdings	total area hectares	area as % of total	holdings	total area hectares	area as % of total	holdings	total area hectares	area as % of total
Under 20 ha	1 974	16 042	9.6	954	8 312	19.8	2 913	21 281	9.6
20 to <100 ha	2 118	101 933	60.7	553	22 572	53.9	2 401	112 187	50.8
100 to <300 ha	291	41 390	24.7	44	6 870	16.4	366	54 593	24.7
300 ha and over	14	8 540	5.1	5	4 144	9.9	53	32 770	14.8
Total	**4 397**	**167 905**	*100.0*	**1 556**	**41 898**	*100.0*	**5 733**	**220 831**	*100.0*
ANALYSIS BY FARM TYPE	holdings	total area hectares	area as % of total	holdings	total area hectares	area as % of total	holdings	total area hectares	area as % of total
Dairying	1 516	89 462	53.3	235	9 860	23.5	1 426	78 945	35.7
Cattle & sheep	1 165	32 158	19.2	561	15 671	37.4	1 859	93 349	42.3
Cropping	514	29 353	17.5	155	8 693	20.7	505	24 846	11.3
Pigs and poultry	231	3 930	2.3	137	1 809	4.3	390	4 948	2.2
Horticulture	226	2 155	1.3	99	811	1.9	617	5 490	2.5
Unclassified	745	10 848	6.:	369	5 054	12.1	936	13 254	6.0
Total	**4 397**	**167 905**	*100.0*	**1 556**	**41 898**	*100.0*	**5 733**	**220 831**	*100.0*

English Region and County Statistics for Main Holdings
June Census 1991

LAND	holdings	hectares	area as % of England
Total agricultural area	**513**	**19 948**	*0.2*
Total cereals	**239**	**8 199**	*0.3*
Wheat	168	3 612	*0.2*
Barley	195	4 237	*0.4*
Other cereals	38	350	*0.4*
Crops mainly for stockfeed			
Peas for harvesting dry	41	402	*0.6*
Field beans	51	542	*0.4*
Other crops for stockfeed	46	230	*0.3*
Other arable crops			
Potatoes	142	934	*0.7*
Sugar beet	24	222	*0.1*
Oilseed rape	45	620	*0.2*
Other arable crops	9	64	*0.1*
Horticultural crops			
Vegetables	140	1 041	*0.8*
Orchards and small fruit	15	60	*0.1*
HNS, bulbs and flowers	37	60	*0.5*
Glasshouse area	54	14	*0.7*
Grassland &all other land			
Grass <5 years old	148	1 993	*0.2*
All other grassland	327	4 233	*0.1*
All other land	283	1 335	*0.3*

LIVESTOCK	holdings	number	number as % of England
Total cattle and calves	**145**	**12 169**	*0.2*
Dairy cows	39	2 997	*0.2*
Beef cows	46	1 019	*0.1*
Total pigs	**35**	**8 866**	*0.1*
Breeding herd	23	806	*0.1*
Total sheep and lambs	**35**	**8 139**	*0.0*
Breeding flock	33	3 458	*0.0*
Total fowls	**61**	**447 262**	*0.5*
Laying birds	53	236 419	*0.9*
Table birds	5	113 905	*0.2*

LABOUR	holdings	number	number as % of England
Total labour force	**426**	**1 570**	*0.4*
Farmers, partners & directors	410	577	*0.3*
Regular whole-time workers	152	379	*0.4*
Regular part-time workers	79	170	*0.4*
Seasonal and casual workers	90	272	*0.4*

ANALYSIS BY TOTAL AREA	holdings	total area hectares	area as % of total
Under 20 ha	263	1 641	*8.2*
20 to <100 ha	203	10 517	*52.7*
100 to <300 ha	44	6 600	*33.1*
300 ha and over	3	1 190	*6.0*
Total	**513**	**19 948**	*100.0*

ANALYSIS BY FARM TYPE	holdings	total area hectares	area as % of total
Dairying	30	2 515	*12.6*
Cattle & sheep	62	2 073	*10.4*
Cropping	222	12 898	*64.7*
Pigs and poultry	35	314	*1.6*
Horticulture	91	1 011	*5.1*
Unclassified	73	1 138	*5.7*
Total	**513**	**19 948**	*100.0*

Merseyside

5-23

Welsh County Statistics for Main Holdings
June Census 1991

	WALES			Clwyd			Dyfed		
LAND	holdings	hectares	area as % of Wales	holdings	hectares	area as % of Wales	holdings	hectares	area as % of Wales
Total agricultural area	29 710	1 492 264	100.0	3 829	180 269	12.1	11 107	455 730	30.5
Total cereals	3 951	53 029	100.0	602	8 081	15.2	1 575	18 398	34.7
Wheat	703	11 780	100.0	123	1 982	16.8	117	1 684	14.3
Barley	3 455	36 143	100.0	523	5 513	15.3	1 453	15 181	42.0
Other cereals	1 036	5 106	100.0	127	586	11.5	318	1 534	30.0
Crops mainly for stockfeed									
Peas for harvesting dry	59	410	100.0	7	77	18.8	20	172	41.9
Field beans	67	588	100.0	17	169	28.8	8	18	3.1
Other crops for stockfeed	1 736	6 930	100.0	231	1 059	15.3	447	1 599	23.1
Other arable crops									
Potatoes	1 634	4 179	100.0	175	298	7.1	649	2 259	54.1
Sugar beet	16	104	100.0	7	67	64.6	3	7	6.8
Oilseed rape	62	1 084	100.0	7	156	14.4	17	200	18.5
Other arable crops	80	659	100.0	14	162	24.6	26	245	37.2
Horticultural crops									
Vegetables	336	799	100.0	39	141	17.7	128	255	31.9
Orchards and small fruit	176	345	100.0	19	21	6.1	53	37	10.7
HNS, bulbs and flowers	123	324	100.0	32	178	55.0	32	67	20.7
Glasshouse area	147	44	100.0	34	8	18.3	40	8	18.3
Grassland &all other land									
Grass <5 years old	10 174	147 867	100.0	1 419	20 651	14.0	3 807	52 970	35.8
All other grassland	28 448	1 221 930	100.0	3 621	143 828	11.8	10 630	359 824	29.4
All other land	13 179	53 975	100.0	1 566	5 376	10.0	5 295	19 671	36.4

	WALES			Clwyd			Dyfed		
LIVESTOCK	holdings	number	number as % of Wales	holdings	number	number as % of Wales		number	number as % of Wales
Total cattle and calves	19 239	1 335 276	100.0	2 486	196 390	14.7	7 408	529 246	39.6
Dairy cows	6 141	317 091	100.0	1 007	59 262	18.7	3 338	174 677	55.1
Beef cows	11 315	206 196	100.0	1 203	20 577	10.0	3 712	51 175	24.8
Total pigs	1 375	102 038	100.0	177	25 626	25.1	581	17 781	17.4
Breeding herd	950	11 749	100.0	129	2 129	18.1	394	2 658	22.6
Total sheep and lambs	17 513	10 782 143	100.0	2 288	1 523 353	14.1	5 211	2 215 082	20.5
Breeding flock	17 155	5 181 020	100.0	2 229	710 842	13.7	5 057	1 068 472	20.6
Total fowls	5 357	6 877 338	100.0	599	1 547 712	22.5	1 966	701 397	10.2
Laying birds	4 999	1 014 271	100.0	557	172 978	17.1	1 843	240 448	23.7
Table birds	259	4 691 734	100.0	27	1 319 506	28.1	91	22 494	0.5

	WALES			Clwyd			Dyfed		
LABOUR	holdings	number	number as % of Wales	holdings	number	number as % of Wales		number	number as % of Wales
Total labour force	25 372	61 020	100.0	3 259	8 309	13.6	9 461	22 576	37.0
Farmers, partners & directors	25 028	31 750	100.0	3 217	4 233	13.3	9 336	11 611	36.6
Regular whole-time workers	3 816	5 730	100.0	651	1 028	17.9	1 349	1 966	34.3
Regular part-time workers	2 765	3 779	100.0	400	543	14.4	886	1 217	32.2
Seasonal and casual workers	4 053	8 461	100.0	583	1 082	12.8	1 320	3 175	37.5

	WALES			Clwyd			Dyfed		
ANALYSIS BY TOTAL AREA	holdings	total area hectares	area as % of total	holdings	total area hectares	area as % of total	holdings	total area hectares	area as % of total
Under 20 ha	11 818	108 018	7.2	1 604	13 379	7.4	4 620	43 310	9.5
20 to <100 ha	14 283	689 557	46.2	1 789	87 997	48.8	5 614	264 317	58.0
100 to <300 ha	3 196	483 639	32.4	389	57 946	32.1	811	116 131	25.5
300 ha and over	413	211 050	14.1	47	20 947	11.6	62	31 973	7.0
Total	29 710	1 492 264	100.0	3 829	180 269	100.0	11 107	455 730	100.0

	WALES			Clwyd			Dyfed		
ANALYSIS BY FARM TYPE	holdings	total area hectares	area as % of total	holdings	total area hectares	area as % of total	holdings	total area hectares	area as % of total
Dairying	5 076	278 673	18.7	771	43 097	23.9	2 975	162 338	35.6
Cattle & sheep	18 441	1 075 467	72.1	2 255	122 359	67.9	5 679	242 327	53.2
Cropping	697	43 118	2.9	89	4 930	2.7	308	16 809	3.7
Pigs and poultry	598	8 955	0.6	90	1 015	0.6	180	2 301	0.5
Horticulture	267	2 519	0.2	57	590	0.3	79	644	0.1
Unclassified	4 631	83 531	5.6	567	8 278	4.6	1 886	31 311	6.9
Total	29 710	1 492 264	100.0	3 829	180 269	100.0	11 107	455 730	100.0

Welsh County Statistics for Main Holdings
June Census 1991

	Gwent			Gwynedd			Mid Glamorgan		
LAND	holdings	hectares	area as % of Wales	holdings	hectares	area as % of Wales	holdings	hectares	area as % of Wales
Total agricultural area	**2 143**	**82 008**	**5.5**	**4 822**	**297 231**	**19.9**	**934**	**49 355**	**3.3**
Total cereals	**302**	**5 992**	**11.3**	**494**	**4 383**	**8.3**	**57**	**1 792**	**3.4**
Wheat	164	2 910	24.7	21	224	1.9	14	460	3.9
Barley	213	2 408	6.7	397	3 488	9.7	51	1 151	3.2
Other cereals	113	675	13.2	159	671	13.1	21	180	3.5
Crops mainly for stockfeed									
Peas for harvesting dry	14	85	20.7	5	16	3.9	*	*	*
Field beans	19	176	30.0	*	*	*	6	76	12.9
Other crops for stockfeed	269	1 295	18.7	197	637	9.2	55	305	4.4
Other arable crops									
Potatoes	172	491	11.7	291	336	8.0	33	104	2.5
Sugar beet	*	*	*	*	*	*	0	0	0.0
Oilseed rape	*	*	*	5	41	3.8	4	161	14.9
Other arable crops	7	50	7.6	9	15	2.3	3	42	6.4
Horticultural crops									
Vegetables	39	74	9.3	30	45	5.6	11	57	7.1
Orchards and small fruit	41	159	46.2	20	33	9.6	5	17	4.9
HNS, bulbs and flowers	12	16	4.9	22	14	4.3	7	7	2.2
Glasshouse area	18	5	11.5	18	3	6.9	6	1	2.3
Grassland &all other land									
Grass <5 years old	678	10 121	6.8	1 446	21 630	14.6	243	3 512	2.4
All other grassland	2 038	59 337	4.9	4 627	262 754	21.5	902	41 318	3.4
All other land	994	3 989	7.4	1 591	7 317	13.6	401	1 964	3.6
LIVESTOCK			number as % of Wales			number as % of Wales			number as % of Wales
Total cattle and calves	**1 306**	**91 803**	**6.9**	**3 007**	**175 056**	**13.1**	**536**	**30 515**	**2.3**
Dairy cows	370	22 736	7.2	579	19 597	6.2	85	4 158	1.3
Beef cows	738	9 447	4.6	1 898	34 458	16.7	369	7 042	3.4
Total pigs	**119**	**28 786**	**28.2**	**219**	**9 257**	**9.1**	**54**	**2 904**	**2.8**
Breeding herd	83	3 152	26.8	157	1 197	10.2	41	367	3.1
Total sheep and lambs	**1 146**	**449 597**	**4.2**	**3 321**	**2 100 701**	**19.5**	**519**	**384 618**	**3.6**
Breeding flock	1 123	224 702	4.3	3 259	1 041 974	20.1	510	195 173	3.8
Total fowls	**387**	**1 217 610**	**17.7**	**879**	**1 911 829**	**27.8**	**173**	**184 726**	**2.7**
Laying birds	354	191 005	18.8	824	69 817	6.9	162	64 859	6.4
Table birds	26	857 637	18.3	26	1 726 320	36.8	11	108 775	2.3
LABOUR			number as % of Wales			number as % of Wales			number as % of Wales
Total labour force	**1 844**	**4 479**	**7.3**	**3 988**	**8 581**	**14.1**	**799**	**1 917**	**3.1**
Farmers, partners & directors	1 811	2 277	7.2	3 938	4 840	15.2	790	997	3.1
Regular whole-time workers	309	478	8.3	525	762	13.3	103	150	2.6
Regular part-time workers	265	387	10.2	391	494	13.1	101	146	3.9
Seasonal and casual workers	350	588	6.9	608	978	11.6	175	339	4.0
ANALYSIS BY TOTAL AREA	holdings	total area hectares	area as % of total	holdings	total area hectares	area as % of total	holdings	total area hectares	area as % of total
Under 20 ha	916	8 799	10.7	2 000	18 430	6.2	367	3 499	7.1
20 to <100 ha	1 064	49 964	60.9	2 028	93 888	31.6	443	21 075	42.7
100 to <300 ha	*	*	*	636	104 324	35.1	106	17 350	35.2
300 ha and over	*	*	*	158	80 589	27.1	18	7 432	15.1
Total	**2 143**	**82 008**	**100.0**	**4 822**	**297 231**	**100.0**	**934**	**49 355**	**100.0**
ANALYSIS BY FARM TYPE	holdings	total area hectares	area as % of total	holdings	total area hectares	area as % of total	holdings	total area hectares	area as % of total
Dairying	316	19 495	23.8	413	18 436	6.2	62	3 886	7.9
Cattle & sheep	1 243	47 683	58.1	3 551	261 068	87.8	617	37 816	76.6
Cropping	86	6 201	7.6	61	2 457	0.8	16	2 602	5.3
Pigs and poultry	86	1 354	1.7	79	790	0.3	24	497	1.0
Horticulture	33	535	0.7	34	225	0.1	12	75	0.2
Unclassified	379	6 740	8.2	684	14 254	4.8	203	4 480	9.1
Total	**2 143**	**82 008**	**100.0**	**4 822**	**297 231**	**100.0**	**934**	**49 355**	**100.0**

Welsh County Statistics for Main Holdings
June Census 1991

	Powys			South Glamorgan			West Glamorgan		
LAND			area as % of Wales			area as % of Wales			area as % of Wales
	holdings	hectares		holdings	hectares		holdings	hectares	
Total agricultural area	5 506	367 815	24.6	495	24 313	6.6	874	35 543	2.4
Total cereals	677	9 614	18.1	118	3 635	37.8	126	1 134	2.1
Wheat	210	3 061	26.0	39	1 321	43.2	15	137	1.2
Barley	596	5 485	15.2	108	2 052	37.4	114	867	2.4
Other cereals	233	1 068	20.9	32	262	24.5	33	130	2.5
Crops mainly for stockfeed									
Peas for harvesting dry	12	60	14.6	0	0	0.0	*	*	*
Field beans	6	55	9.4	7	86	156.4	*	*	*
Other crops for stockfeed	387	1 437	20.7	34	324	22.5	116	274	4.0
Other arable crops									
Potatoes	160	294	7.0	32	108	36.7	122	290	6.9
Sugar beet	3	13	12.5	0	0	0.0	0	0	0.0
Oilseed rape	12	176	16.2	4	109	61.9	*	*	*
Other arable crops	14	76	11.5	7	70	92.1	0	0	0.0
Horticultural crops									
Vegetables	18	11	1.4	11	21	190.9	60	194	24.3
Orchards and small fruit	22	32	9.3	*	*	*	*	*	*
HNS, bulbs and flowers	11	36	11.1	*	*	*	*	*	*
Glasshouse area	8	2	4.6	11	15	750.0	12	2	4.6
Grassland &all other land									
Grass <5 years old	2 148	32 727	22.1	153	2 941	9.0	280	3 317	2.2
All other grassland	5 340	310 565	25.4	453	15 970	5.1	837	28 333	2.3
All other land	2 671	12 716	23.6	240	1 011	8.0	421	1 932	3.6
LIVESTOCK			number as % of Wales			number as % of Wales			number as % of Wales
Total cattle and calves	3 686	251 787	18.9	296	30 988	12.3	514	29 491	2.2
Dairy cows	549	25 188	7.9	115	7 860	31.2	98	3 613	1.1
Beef cows	2 923	74 502	36.1	128	2 350	3.2	344	6 645	3.2
Total pigs	157	10 307	10.1	22	5 451	52.9	46	1 926	1.9
Breeding herd	98	1 378	11.7	16	612	44.4	32	256	2.2
Total sheep and lambs	4 456	3 835 803	35.6	189	73 000	1.9	383	199 989	1.9
Breeding flock	4 423	1 805 595	34.9	183	34 339	1.9	371	99 923	1.9
Total fowls	1 141	1 093 582	15.9	67	96 318	8.8	145	124 164	1.8
Laying birds	1 072	140 577	13.9	55	16 725	11.9	132	117 862	11.6
Table birds	63	576 889	12.3	8	77 945	13.5	7	2 168	0.0
LABOUR			number as % of Wales			number as % of Wales			number as % of Wales
Total labour force	4 887	12 034	19.7	411	1 305	10.8	723	1 819	3.0
Farmers, partners & directors	4 824	6 342	20.0	402	541	8.5	710	909	2.9
Regular whole-time workers	686	994	17.3	104	208	20.9	89	144	2.5
Regular part-time workers	574	746	19.7	62	116	15.5	86	130	3.4
Seasonal and casual workers	783	1 689	20.0	76	258	15.3	158	352	4.2
ANALYSIS BY TOTAL AREA	holdings	total area hectares	area as % of total	holdings	total area hectares	area as % of total	holdings	total area hectares	area as % of total
Under 20 ha	1 712	14 779	4.0	201	1 923	7.9	398	3 900	11.0
20 to <100 ha	2 719	143 544	39.0	229	11 148	45.9	397	17 624	49.6
100 to <300 ha	957	144 058	39.2	*	*	*	75	12 072	34.0
300 ha and over	118	65 434	17.8	*	*	*	4	1 948	5.5
Total	5 506	367 815	100.0	495	24 313	100.0	874	35 543	100.0
ANALYSIS BY FARM TYPE	holdings	total area hectares	area as % of total	holdings	total area hectares	area as % of total	holdings	total area hectares	area as % of total
Dairying	353	18 942	5.1	105	8 257	34.0	81	4 223	11.9
Cattle & sheep	4 355	328 248	89.2	219	10 242	42.1	522	25 725	72.4
Cropping	59	4 928	1.3	35	3 288	13.5	43	1 901	5.3
Pigs and poultry	100	2 449	0.7	17	278	1.1	22	271	0.8
Horticulture	19	118	0.0	13	86	0.4	20	246	0.7
Unclassified	620	13 130	3.6	106	2 161	8.9	186	3 179	8.9
Total	5 506	367 815	100.0	495	24 313	100.0	874	35 543	100.0

Scottish Region Statistics for Main Holdings
June Census 1991

	SCOTLAND			Shetland			Orkney		
LAND	holdings	hectares	area as % of Scotland	holdings	hectares	area as % of Scotland	holdings	hectares	area as % of Scotland
Total agricultural area	30 902	5 269 574	100.0	1 291	63 171	1.2	1 303	79 709	1.5
Total cereals	13 254	467 471	100.0	253	197	0.0	590	3 449	0.7
Wheat	4 080	109 675	100.0	*	*	*	*	*	*
Barley	11 382	329 115	100.0	*	*	*	*	*	*
Other cereals	3 669	28 681	100.0	232	152	0.5	135	272	0.9
Crops mainly for stockfeed									
Peas for harvesting dry	263	3 369	100.0	*	*	*	*	*	*
Other crops for stockfeed	7 246	34 477	100.0	392	198	0.6	202	500	1.5
Other arable crops									
Potatoes	5 597	26 370	100.0	409	102	0.4	299	190	0.7
Oilseed rape	2 279	49 895	100.0	*	*	*	2	5	0.0
Other arable crops	247	2 337	100.0	9	2	0.1	6	11	0.5
Horticultural crops									
Vegetables	1 213	11 473	100.0	61	14	0.1	29	22	0.2
Orchards and small fruit	586	2 990	100.0	*	*	*	*	*	*
HNS, bulbs and flowers	209	820	100.0	*	*	*	*	*	*
Glasshouse area	314	47	100.0	*	*	*	*	*	*
Grassland & all other land									
Grass <5 years old	18 535	395 552	100.0	574	2 506	0.6	946	12 378	3.1
All other grassland	26 772	4 105 736	100.0	1 281	59 822	1.5	1 267	59 537	1.5
All other land	23 300	169 039	100.0	557	328	0.2	1 076	3 617	2.1
LIVESTOCK	holdings	number	number as % of Scotland	holdings	number	number as % of Scotland	holdings	number	number as % of Scotland
Total cattle and calves	17 655	2 107 866	100.0	270	5 594	0.3	922	100 077	4.7
Dairy cows	3 118	240 668	100.0	24	595	0.2	69	3 667	1.5
Beef cows	11 876	486 340	100.0	220	1 911	0.4	813	30 391	6.2
Total pigs	902	493 025	100.0	*	*	*	60	833	0.2
Breeding herd	546	51 375	100.0	*	*	*	21	98	0.2
Total sheep and lambs	15 954	9 757 429	100.0	1 203	356 456	3.7	547	142 191	1.5
Breeding flock	15 606	4 692 472	100.0	1 199	189 736	4.0	532	64 591	1.4
Total fowls	4 530	13 604 929	100.0	188	2 555	0.0	321	8 231	0.1
Laying birds	4 213	2 594 992	100.0	184	2 359	0.1	306	6 187	0.2
Table birds	226	9 057 390	100.0	4	11	0.0	38	1 468	0.0
LABOUR	holdings	number	number as % of Scotland	holdings	number	number as % of Scotland	holdings	number	number as % of Scotland
Total labour force	25 516	61 104	100.0	1 044	1 734	2.8	1 092	2 199	3.6
Farmers	23 398	23 398	100.0	1 009	1 009	4.3	1 051	1 051	4.5
Regular whole-time workers	9 491	19 339	100.0	37	60	0.3	249	376	1.9
Regular part-time workers	3 553	5 195	100.0	85	127	2.4	116	147	2.8
Seasonal and casual workers	1 642	2 779	100.0	31	56	2.0	61	84	3.0
		total area hectares	area as % of total		total area hectares	area as % of total		total area hectares	area as % of total
ANALYSIS BY TOTAL AREA	holdings			holdings			holdings		
Under 20 ha	9 749	73 554	1.4	732	6 124	9.7	407	4 271	5.4
20 to <100 ha	12 016	638 601	12.1	441	19 226	30.4	714	32 354	40.6
100 to <300 ha	6 426	1 057 482	20.1	80	13 057	20.7	152	23 157	29.1
300 ha and over	2 711	3 499 937	66.4	38	24 764	39.2	30	19 926	25.0
Total	30 902	5 269 574	100.0	1 291	63 171	100.0	1 303	79 709	100.0
		total area hectares	area as % of total		total area hectares	area as % of total		total area hectares	area as % of total
ANALYSIS BY FARM TYPE	holdings			holdings			holdings		
Dairying	2 515	282 744	5.4	*	*	*	43	5 234	6.6
Cattle & sheep	6 885	2 997 477	56.9	90	26 372	42.3	417	49 074	61.6
Cropping	5 014	693 216	13.2	*	*	*	17	1 143	1.4
Pigs and poultry	167	4 485	0.1	*	*	*	0	0	0.0
Horticulture	94	2 577	0.0	*	*	*	0	0	0.0
Unclassified	16 227	1 289 074	24.5	1 194	35 919	57.7	826	24 258	30.4
Total	30 902	5 269 573	100.0	1 284	62 291	100.0	1 303	79 709	100.0

Scottish Region Statistics for Main Holdings
June Census 1991

	Western Isles			Highland			Grampian		
LAND	holdings	hectares	area as % of Scotland	holdings	hectares	area as % of Scotland	holdings	hectares	area as % of Scotland
Total agricultural area	1 589	51 115	1.0	4 486	1 668 339	31.7	6 185	621 374	11.8
Total cereals	286	593	0.1	1 377	31 959	6.8	4 097	139 013	29.7
Wheat	*	*	*	170	4 074	3.7	1 401	25 187	23.0
Barley	*	*	*	988	23 625	7.2	3 652	103 347	31.4
Other cereals	277	568	2.0	657	4 261	14.9	1 108	10 479	36.5
Crops mainly for stockfeed									
Peas for harvesting dry	*	*	*	37	484	14.4	23	155	4.6
Other crops for stockfeed	95	35	0.1	1 065	3 995	11.6	2 328	11 024	32.0
Other arable crops									
Potatoes	513	92	0.3	770	930	3.5	1 275	5 621	21.3
Oilseed rape	*	*	*	104	2 052	4.1	776	16 756	33.6
Other arable crops	4	1	0.0	16	131	5.6	46	429	18.4
Horticultural crops									
Vegetables	101	17	0.1	67	44	0.4	235	1 707	14.9
Orchards and small fruit	0	0	0.0	35	83	2.8	83	182	6.1
HNS, bulbs and flowers	2	1	0.1	8	7	0.8	49	416	50.8
Glasshouse area	2	0	0.4	26	2	3.2	48	6	12.5
Grassland & all other land									
Grass <5 years old	388	1 244	0.3	2 590	40 666	10.3	4 548	114 887	29.0
All other grassland	1 462	47 121	1.1	4 042	1 545 138	37.6	4 636	302 390	7.4
All other land	644	2 011	1.2	2 764	42 849	25.3	4 807	28 788	17.0
LIVESTOCK	holdings	number	number as % of Scotland	holdings	number	number as % of Scotland	holdings	number	number as % of Scotland
Total cattle and calves	430	6 338	0.3	2 239	156 055	7.4	3 852	429 404	20.4
Dairy cows	35	234	0.1	92	4 794	2.0	286	23 805	9.9
Beef cows	375	2 683	0.6	1 957	59 487	12.2	2 064	91 140	18.7
Total pigs	*	*	*	80	16 737	3.4	347	271 933	55.2
Breeding herd	*	*	*	44	1 901	3.7	224	28 688	55.8
Total sheep and lambs	1 373	224 994	2.3	2 898	1 371 546	14.1	2 178	824 174	8.4
Breeding flock	1 368	124 907	2.7	2 862	725 682	15.5	2 068	362 757	7.7
Total fowls	311	3 767	0.0	746	210 585	1.5	826	2 317 844	17.0
Laying birds	289	3 331	0.1	716	56 385	2.2	759	398 935	15.4
Table birds	*	*	*	25	142 487	1.6	41	1 791 808	19.8
LABOUR	holdings	number	number as % of Scotland	holdings	number	number as % of Scotland	holdings	number	number as % of Scotland
Total labour force	1 123	1 707	2.8	3 539	6 546	10.7	5 108	11 595	19.0
Farmers	1 060	1 060	4.5	3 246	3 246	13.9	4 737	4 737	20.2
Regular whole-time workers	62	99	0.5	691	1 187	6.1	1 684	3 608	18.7
Regular part-time workers	106	167	3.2	375	481	9.3	597	856	16.5
Seasonal and casual workers	28	53	1.9	175	289	10.4	264	447	16.1
ANALYSIS BY TOTAL AREA	holdings	total area hectares	area as % of total	holdings	total area hectares	area as % of total	holdings	total area hectares	area as % of total
Under 20 ha	1 409	8 692	17.0	2 025	14 658	0.9	1 670	13 700	2.2
20 to <100 ha	129	4 765	9.3	1 328	62 442	3.7	3 035	159 422	25.7
100 to <300 ha	29	4 687	9.2	572	96 954	5.8	1 240	198 200	31.9
300 ha and over	22	32 970	64.5	561	1 494 285	89.6	240	250 053	40.2
Total	1 589	51 115	100.0	4 486	1 668 339	100.0	6 185	621 374	100.0
ANALYSIS BY FARM TYPE	holdings	total area hectares	area as % of total	holdings	total area hectares	area as % of total	holdings	total area hectares	area as % of total
Dairying	*	*	*	40	6 481	0.4	174	25 430	4.1
Cattle & sheep	27	10 810	21.2	855	908 439	54.5	1 195	219 671	35.4
Cropping	*	*	*	325	50 708	3.0	1 741	209 992	33.8
Pigs and poultry	*	*	*	7	165	0.0	34	1 714	0.3
Horticulture	*	*	*	6	183	0.0	13	710	0.1
Unclassified	1 560	40 154	78.8	3 253	702 363	42.1	3 028	163 856	26.4
Total	1 587	50 964	100.0	4 486	1 668 339	100.0	6 185	621 373	100.0

Scottish Region Statistics for Main Holdings
June Census 1991

	Tayside			Fife			Lothian		
LAND	holdings	hectares	area as % of Scotland	holdings	hectares	area as % of Scotland	holdings	hectares	area as % of Scotland
Total agricultural area	**2 772**	**621 503**	*11.8*	**995**	**97 448**	*15.7*	**1 089**	**125 184**	*2.4*
Total cereals	**1 805**	**92 268**	*19.7*	**685**	**42 599**	*46.2*	**623**	**41 668**	*8.9*
Wheat	882	21 437	*19.5*	485	15 351	*71.6*	389	18 492	*16.9*
Barley	1 719	66 314	*20.1*	639	26 056	*39.3*	577	22 397	*6.8*
Other cereals	383	4 517	*15.7*	117	1 192	*26.4*	68	779	*2.7*
Crops mainly for stockfeed									
Peas for harvesting dry	37	439	*13.0*	18	236	*53.9*	66	1 010	*30.0*
Other crops for stockfeed	893	5 081	*14.7*	234	1 289	*25.4*	183	1 354	*3.9*
Other arable crops									
Potatoes	1 123	11 232	*42.6*	289	2 896	*25.8*	163	2 015	*7.6*
Oilseed rape	589	12 364	*24.8*	311	6 781	*54.8*	187	4 302	*8.6*
Other arable crops	47	411	*17.6*	21	199	*48.4*	20	368	*15.8*
Horticultural crops									
Vegetables	295	3 829	*33.4*	134	3 213	*83.9*	89	868	*7.6*
Orchards and small fruit	300	2 324	*77.7*	30	167	*7.2*	27	61	*2.1*
HNS, bulbs and flowers	48	207	*25.2*	16	20	*9.8*	25	86	*10.4*
Glasshouse area	23	4	*7.8*	21	2	*43.2*	32	5	*11.4*
Grassland & all other land									
Grass <5 years old	1 808	38 602	*9.8*	567	12 059	*31.2*	494	12 341	*3.1*
All other grassland	1 965	434 963	*10.6*	757	22 258	*5.1*	824	54 330	*1.3*
All other land	2 316	19 780	*11.7*	851	5 730	*29.0*	895	6 776	*4.0*
LIVESTOCK	holdings	number	number as % of Scotland	holdings	number	number as % of Scotland	holdings	number	number as % of Scotland
Total cattle and calves	**1 374**	**134 916**	*6.4*	**512**	**62 920**	*46.6*	**465**	**57 053**	*2.7*
Dairy cows	84	6 625	*2.8*	88	7 148	*107.9*	91	5 490	*2.3*
Beef cows	847	40 584	*8.3*	286	13 276	*32.7*	258	12 889	*2.7*
Total pigs	**80**	**66 295**	*13.4*	**40**	**26 166**	*39.5*	**42**	**45 768**	*9.3*
Breeding herd	61	7 378	*14.4*	30	3 137	*42.5*	33	4 961	*9.7*
Total sheep and lambs	**1 011**	**884 961**	*9.1*	**269**	**129 430**	*14.6*	**334**	**278 219**	*2.9*
Breeding flock	984	432 680	*9.2*	260	54 755	*12.7*	326	125 188	*2.7*
Total fowls	**314**	**3 101 469**	*22.8*	**122**	**2 624 702**	*84.6*	**132**	**1 693 617**	*12.4*
Laying birds	271	201 491	*7.8*	101	1 184 551	*587.9*	106	91 678	*3.5*
Table birds	34	2 679 244	*29.6*	12	883 731	*33.0*	16	1 450 014	*16.0*
LABOUR	holdings	number	number as % of Scotland	holdings	number	number as % of Scotland	holdings	number	number as % of Scotland
Total labour force	**2 350**	**6 335**	*10.4*	**829**	**2 774**	*43.8*	**863**	**2 857**	*4.7*
Farmers	2 081	2 081	*8.9*	759	759	*36.5*	747	747	*3.2*
Regular whole-time workers	1 147	2 491	*12.9*	454	1 123	*45.1*	477	1 372	*7.1*
Regular part-time workers	372	524	*10.1*	141	395	*75.4*	167	289	*5.6*
Seasonal and casual workers	165	406	*14.6*	53	169	*41.6*	68	148	*5.3*
ANALYSIS BY TOTAL AREA	holdings	total area hectares	area as % of total	holdings	total area hectares	area as % of total	holdings	total area hectares	area as % of total
Under 20 ha	626	4 230	*0.7*	255	1 337	*1.4*	361	2 229	*1.8*
20 to <100 ha	1 050	60 573	*9.7*	354	21 159	*21.7*	345	18 802	*15.0*
100 to <300 ha	779	129 249	*20.8*	346	56 885	*58.4*	294	51 018	*40.8*
300 ha and over	317	427 451	*68.8*	40	18 068	*18.5*	89	53 135	*42.4*
Total	**2 772**	**621 503**	*100.0*	**995**	**97 448**	*100.0*	**1 089**	**125 184**	*100.0*
ANALYSIS BY FARM TYPE	holdings	total area hectares	area as % of total	holdings	total area hectares	area as % of total	holdings	total area hectares	area as % of total
Dairying	51	7 040	*1.1*	68	9 248	*9.5*	57	6 700	*5.4*
Cattle & sheep	489	387 682	*62.4*	138	17 800	*18.3*	144	43 701	*34.9*
Cropping	1 240	163 333	*26.3*	433	62 952	*64.6*	382	61 562	*49.2*
Pigs and poultry	31	422	*0.1*	23	163	*0.2*	28	477	*0.4*
Horticulture	42	1 127	*0.2*	3	45	*0.0*	8	225	*0.2*
Unclassified	919	61 900	*10.0*	330	7 240	*7.4*	470	12 518	*10.0*
Total	**2 772**	**621 504**	*100.0*	**995**	**97 448**	*100.0*	**1 089**	**125 183**	*100.0*

Scottish Region Statistics for Main Holdings
June Census 1991

	Borders			Central			Strathclyde		
LAND	holdings	hectares	area as % of Scotland	holdings	hectares	area as % of Scotland	holdings	hectares	area as % of Scotland
Total agricultural area	1 676	385 365	7.3	1 057	197 210	3.7	5 551	912 842	17.3
Total cereals	867	60 948	13.0	385	12 899	2.8	1 382	23 093	4.9
Wheat	474	20 537	18.7	138	2 502	2.3	76	1 061	1.0
Barley	834	37 211	11.3	345	9 208	2.8	1 247	20 903	6.4
Other cereals	198	3 200	11.2	121	1 189	4.1	228	1 128	3.9
Crops mainly for stockfeed									
Peas for harvesting dry	69	1 019	30.2	5	24	0.7	*	*	*
Other crops for stockfeed	615	5 614	16.3	167	751	2.2	620	2 447	7.1
Other arable crops									
Potatoes	153	1 685	6.4	76	162	0.6	403	1 021	3.9
Oilseed rape	209	6 305	12.6	60	902	1.8	*	*	*
Other arable crops	42	571	24.4	9	26	1.1	13	32	1.4
Horticultural crops									
Vegetables	92	1 518	13.2	10	66	0.6	74	159	1.4
Orchards and small fruit	19	30	1.0	10	24	0.8	65	103	3.5
HNS, bulbs and flowers	8	13	1.6	9	5	0.6	33	58	7.0
Glasshouse area	13	1	1.5	20	3	5.3	100	24	50.2
Grassland & all other land									
Grass <5 years old	1 100	40 069	10.1	569	11 291	2.9	3 066	57 948	14.6
All other grassland	1 527	251 195	6.1	960	165 073	4.0	5 244	804 645	19.6
All other land	1 460	16 399	9.7	855	5 984	3.5	4 525	23 169	13.7
LIVESTOCK	holdings	number	number as % of Scotland	holdings	number	number as % of Scotland	holdings	number	number as % of Scotland
Total cattle and calves	996	147 622	7.0	630	63 132	3.0	3 662	503 958	23.9
Dairy cows	71	5 362	2.2	94	6 263	2.6	1 383	103 898	43.2
Beef cows	752	48 077	9.9	399	14 085	2.9	2 387	86 584	17.8
Total pigs	38	17 352	3.5	19	3 395	0.7	114	17 792	3.6
Breeding herd	27	1 732	3.4	10	306	0.6	57	1 559	3.0
Total sheep and lambs	1 129	1 498 237	15.4	516	461 885	4.7	2 744	2 007 039	20.6
Breeding flock	1 105	681 309	14.5	495	229 833	4.9	2 685	983 090	21.0
Total fowls	264	1 065 948	7.8	150	746 615	5.5	783	1 038 920	7.6
Laying birds	252	134 090	5.2	138	56 250	2.2	740	357 327	13.8
Table birds	9	722 206	8.0	*	*	*	31	293 293	3.2
LABOUR	holdings	number	number as % of Scotland	holdings	number	number as % of Scotland	holdings	number	number as % of Scotland
Total labour force	1 474	4 309	7.1	870	2 132	3.5	4 632	11 603	19.0
Farmers	1 321	1 321	5.6	783	783	3.3	4 232	4 232	18.1
Regular whole-time workers	852	1 911	9.9	363	650	3.4	2 018	3 561	18.4
Regular part-time workers	273	378	7.3	142	213	4.1	737	1 030	19.8
Seasonal and casual workers	117	138	5.0	67	115	4.1	385	579	20.8
ANALYSIS BY TOTAL AREA	holdings	total area hectares	area as % of total	holdings	total area hectares	area as % of total	holdings	total area hectares	area as % of total
Under 20 ha	320	2 716	0.7	241	2 018	1.0	1 263	10 033	1.1
20 to <100 ha	405	22 164	5.8	460	25 595	13.0	2 598	145 554	15.9
100 to <300 ha	553	102 680	26.6	234	36 994	18.8	1 156	183 674	20.1
300 ha and over	398	257 805	66.9	122	132 602	67.2	534	573 580	62.8
Total	1 676	385 365	100.0	1 057	197 210	100.0	5 551	912 842	100.0
ANALYSIS BY FARM TYPE	holdings	total area hectares	area as % of total	holdings	total area hectares	area as % of total	holdings	total area hectares	area as % of total
Dairying	37	7 622	2.0	71	8 354	4.2	1 236	120 522	13.2
Cattle & sheep	634	261 616	68.0	307	144 743	73.5	1 456	620 875	68.0
Cropping	436	94 074	24.5	166	18 472	9.4	169	16 846	1.8
Pigs and poultry	*	*	*	*	*	*	23	497	0.1
Horticulture	*	*	*	*	*	*	16	234	0.0
Unclassified	560	21 389	5.6	506	25 395	12.9	2 651	153 867	16.9
Total	1 667	384 701	100.0	1 050	196 964	100.0	5 551	912 841	100.0

Dumfries and Galloway			
LAND	holdings	hectares	area as % of Scotland
Total agricultural area	**2 908**	**446 314**	**8.5**
Total cereals	**904**	**18 785**	**4.0**
Wheat	63	1 031	0.9
Barley	852	16 812	5.1
Other cereals	145	943	3.3
Crops mainly for stockfeed			
Peas for harvesting dry	0	0	0.0
Other crops for stockfeed	452	2 191	6.4
Other arable crops			
Potatoes	124	423	1.6
Oilseed rape	16	283	0.6
Other arable crops	14	156	6.7
Horticultural crops			
Vegetables	26	15	0.1
Orchards and small fruit	14	15	0.5
HNS, bulbs and flowers	10	8	1.0
Glasshouse area	18	2	3.6
Grassland & all other land			
Grass <5 years old	1 885	51 562	13.0
All other grassland	2 807	359 265	8.8
All other land	2 550	13 609	8.1

LIVESTOCK	holdings	number	number as % of Scotland
Total cattle and calves	**2 303**	**440 797**	**20.9**
Dairy cows	801	72 787	30.2
Beef cows	1 518	85 233	17.5
Total pigs	**73**	**26 727**	**5.4**
Breeding herd	37	1 609	3.1
Total sheep and lambs	**1 752**	**1 578 297**	**16.2**
Breeding flock	1 722	717 944	15.3
Total fowls	**373**	**790 676**	**5.8**
Laying birds	351	102 408	3.9
Table birds	11	510 146	5.6

LABOUR	holdings	number	number as % of Scotland
Total labour force	**2 592**	**7 313**	**12.0**
Farmers	2 372	2 372	10.1
Regular whole-time workers	1 457	2 901	15.0
Regular part-time workers	442	588	11.3
Seasonal and casual workers	228	295	10.6

ANALYSIS BY TOTAL AREA	holdings	total area hectares	area as % of total
Under 20 ha	440	3 545	0.8
20 to <100 ha	1 157	66 545	14.9
100 to <300 ha	991	160 926	36.1
300 ha and over	320	215 299	48.2
Total	**2 908**	**446 314**	**100.0**

ANALYSIS BY FARM TYPE	holdings	total area hectares	area as % of total
Dairying	731	85 293	19.1
Cattle & sheep	1 133	306 694	68.7
Cropping	103	13 923	3.1
Pigs and poultry	8	166	0.0
Horticulture	3	24	0.0
Unclassified	930	40 215	9.0
Total	**2 908**	**446 315**	**100.0**

Northern Ireland County Statistics for Main Holdings
June Census 1991

	NORTHERN IRELAND			Antrim			Armagh		
LAND	holdings	hectares	area as % of N.Ireland	holdings	hectares	area as % of N.Ireland	holdings	hectares	area as % of N.Ireland
Total agricultural area	29 363	1 009 555	100.0	5 265	216 881	21.5	3 995	91 085	9.0
Total cereals	5 500	46 028	100.0	1 351	9 447	20.5	395	2 869	6.2
Wheat	599	5 887	100.0	88	603	10.2	35	343	5.8
Barley	5 001	37 003	100.0	1 275	8 385	22.7	356	2 236	6.0
Other	964	3 138	100.0	161	459	14.6	67	290	9.2
Total crops mainly for stockfeed	1 285	4 665	100.0	211	655	14.0	136	596	12.8
Other arable crops									
Potatoes	2 763	10 759	100.0	726	3 306	30.7	220	367	3.4
Oilseed rape	126	1 151	100.0	11	67	5.8	6	67	5.8
Horticultural crops									
Vegetables	336	1 317	100.0	48	156	11.8	77	215	16.3
Orchards and small fruit	694	1 878	100.0	33	27	1.4	499	1 646	87.6
HNS, bulbs and flowers	79	142	100.0	19	24	16.9	17	26	18.3
Glasshouse area	82	74	100.0	18	15	20.3	25	12	16.2
Grassland & all other land									
Grass <5 years old	17 058	177 852	100.0	3 051	34 294	19.3	2 173	18 708	10.5
All other grassland	27 292	736 347	100.0	4 899	162 228	22.0	3 661	65 168	8.9
All other land	9 130	29 342	100.0	1 878	6 662	22.7	973	1 411	4.8
LIVESTOCK	holdings	number	number as % of N.Ireland	holdings	number	number as % of N.Ireland	holdings	number	number as % of N.Ireland
Total cattle and calves	25 610	1 533 226	100.0	4 412	313 346	20.4	3 484	178 163	11.6
Dairy cows	6 684	274 058	100.0	1 440	64 583	23.6	846	29 948	10.9
Beef cows	16 602	254 290	100.0	2 515	48 258	19.0	2 084	24 082	9.5
Total pigs	2 586	588 349	100.0	484	131 386	22.3	523	78 128	13.3
Breeding herd	2 333	59 112	100.0	428	13 448	22.8	481	8 229	13.9
Total sheep and lambs	11 350	2 574 194	100.0	2 515	693 639	26.9	1 027	134 252	5.2
Breeding flock	10 859	1 203 750	100.0	2 414	319 780	26.6	950	68 848	5.7
Total fowls	3 038	11 000 196	100.0	656	3 633 125	33.0	318	1 107 522	10.1
Laying birds	2 221	2 748 174	100.0	418	820 967	29.9	238	379 014	13.8
Table birds	299	6 341 820	100.0	110	2 437 920	38.4	33	564 938	8.9
LABOUR	holdings	number	number as % of N.Ireland	holdings	number	number as % of N.Ireland	holdings	number	number as % of N.Ireland
Total labour force	28 680	57 212	100.0	5 214	11 299	19.7	3 778	7 203	12.6
Farmers, partners & directors	27 041	34 291	100.0	4 932	6 463	18.8	3 561	4 357	12.7
Regular whole-time workers	3 505	4 883	100.0	730	1 035	21.2	403	584	12.0
Regular part-time workers	4 035	5 362	100.0	813	1 053	19.6	487	661	12.3
Seasonal and casual workers	4 914	8 114	100.0	1 053	1 704	21.0	514	1 070	13.2
ANALYSIS BY TOTAL AREA	holdings	total area hectares	area as % of total	holdings	total area hectares	area as % of total	holdings	total area hectares	area as % of total
Under 20 ha	13 016	143 393	14.2	1 983	21 261	9.8	2 383	25 243	27.7
20 to <100 ha	14 991	638 808	63.3	2 930	132 603	61.1	1 561	58 906	64.7
100 to <300 ha	1 270	185 353	18.4	319	46 259	21.3	51	6 936	7.6
300 ha and over	86	42 003	4.2	33	16 759	7.7	0	0	0.0
Total	29 363	1 009 555	100.0	5 265	216 882	100.0	3 995	91 085	100.0
ANALYSIS BY FARM TYPE	holdings	total area hectares	area as % of total	holdings	total area hectares	area as % of total	holdings	total area hectares	area as % of total
Dairying	6 358	305 844	30.3	1 353	73 772	34.0	817	29 787	32.7
Beef cattle & sheep	13 879	551 227	54.6	2 282	116 352	53.6	1 658	41 357	45.4
Pigs and/or poultry	445	6 453	0.6	107	1 627	0.8	80	1 038	1.1
Mixed livestock	344	26 406	2.6	80	3 683	1.7	60	2 617	2.9
Crops and livestock	1 234	10 312	1.0	255	2 649	1.2	147	1 335	1.5
Cropping	1 133	42 332	4.2	281	9 661	4.5	81	2 699	3.0
Horticulture	504	8 399	0.8	50	608	0.3	242	3 453	3.8
Unclassified	5 466	58 582	5.8	857	8 529	3.9	910	8 799	9.7
Total	29 363	1 009 555	100.0	5 265	216 881	100.0	3 995	91 085	100.0

Northern Ireland County Statistics for Main Holdings
June Census 1991

	Down			Fermanagh			Londonderry		
LAND	holdings	hectares	area as % of N.Ireland	holdings	hectares	area as % of N.Ireland	holdings	hectares	area as % of N.Ireland
Total agricultural area	**5 631**	**175 526**	**17.4**	**3 631**	**126 421**	**12.5**	**3 832**	**153 979**	**15.3**
Total cereals	**1 995**	**18 229**	**39.6**	**19**	**199**	**0.4**	**1 163**	**11 056**	**24.0**
Wheat	346	3 591	61.0	0	0	0.0	103	1 145	19.4
Barley	1 752	13 180	35.6	3	19	0.1	1 081	9 446	25.5
Other	449	1 458	46.5	23	180	5.7	190	465	14.8
Total crops mainly for stockfeed	449	1 427	30.6	49	292	6.3	208	715	15.3
Other arable crops									
Potatoes	713	3 044	28.3	78	17	0.2	570	3 171	29.5
Oilseed rape	63	563	48.9	*	*	*	35	392	34.1
Horticultural crops									
Vegetables	137	712	54.1	7	18	1.4	35	100	7.6
Orchards and small fruit	47	37	2.0	36	69	3.7	13	8	0.4
HNS, bulbs and flowers	25	71	50.0	*	*	*	8	9	6.3
Glasshouse area	26	11	14.9	0	0	0.0	7	21	28.4
Grassland & all other land									
Grass <5 years old	3 429	36 252	20.4	1 390	15 007	8.4	2 615	30 549	17.2
All other grassland	5 057	110 913	15.1	3 494	108 033	14.7	3 499	103 304	14.0
All other land	1 675	4 267	14.5	940	2 778	9.5	1 413	4 654	15.9

LIVESTOCK	holdings	number	number as % of N.Ireland	holdings	number	number as % of N.Ireland	holdings	number	number as % of N.Ireland
Total cattle and calves	**4 617**	**305 011**	**19.9**	**3 523**	**173 474**	**11.3**	**3 201**	**202 804**	**13.2**
Dairy cows	1 054	59 469	21.7	932	24 359	8.9	738	32 621	11.9
Beef cows	2 602	33 397	13.1	2 889	50 438	19.8	2 133	34 366	13.5
Total pigs	**496**	**110 665**	**18.8**	**127**	**13 706**	**2.3**	**352**	**114 410**	**19.4**
Breeding herd	458	11 699	19.8	116	1 732	2.9	317	9 410	15.9
Total sheep and lambs	**2 414**	**483 266**	**18.8**	**741**	**122 020**	**4.7**	**1 958**	**557 368**	**21.7**
Breeding flock	2 344	229 307	19.0	693	58 002	4.8	1 878	256 697	21.3
Total fowls	**596**	**2 041 257**	**18.6**	**397**	**123 652**	**1.1**	**355**	**999 446**	**9.1**
Laying birds	443	585 038	21.3	314	96 075	3.5	278	247 188	9.0
Table birds	41	1 069 135	16.9	27	552	0.0	25	616 120	9.7

LABOUR	holdings	number	number as % of N.Ireland	holdings	number	number as % of N.Ireland	holdings	number	number as % of N.Ireland
Total labour force	**5 427**	**11 785**	**20.6**	**3 559**	**6 146**	**10.7**	**3 794**	**7 954**	**13.9**
Farmers, partners & directors	5 104	6 741	19.7	3 370	4 052	11.8	3 586	4 595	13.4
Regular whole-time workers	778	1 135	23.2	340	426	8.7	539	701	14.4
Regular part-time workers	838	1 211	22.6	416	496	9.3	566	733	13.7
Seasonal and casual workers	993	1 745	21.5	483	627	7.7	813	1 362	16.8

ANALYSIS BY TOTAL AREA	holdings	total area hectares	area as % of total	holdings	total area hectares	area as % of total	holdings	total area hectares	area as % of total
Under 20 ha	2 744	29 601	16.9	1 456	17 189	13.6	1 534	16 710	10.9
20 to <100 ha	2 669	112 496	64.1	2 019	84 525	66.9	2 016	89 402	58.1
100 to <300 ha	208	29 218	16.6	150	22 149	17.5	264	39 897	25.9
300 ha and over	10	4 211	2.4	6	2 558	2.0	18	7 971	5.2
Total	**5 631**	**175 526**	**100.0**	**3 631**	**126 421**	**100.0**	**3 832**	**153 980**	**100.0**

ANALYSIS BY FARM TYPE	holdings	total area hectares	area as % of total	holdings	total area hectares	area as % of total	holdings	total area hectares	area as % of total
Dairying	998	56 834	32.4	898	37 614	29.8	678	35 796	23.2
Beef cattle & sheep	2 402	73 206	41.7	2 002	80 164	63.4	1 916	92 983	60.4
Pigs and/or poultry	99	1 502	0.9	9	128	0.1	54	829	0.5
Mixed livestock	78	12 878	7.3	7	0	0.0	42	5 581	3.6
Crops and livestock	441	2 429	1.4	5	191	0.2	270	1 045	0.7
Cropping	423	16 293	9.3	3	28	0.0	243	10 098	6.6
Horticulture	122	2 226	1.3	8	144	0.1	24	662	0.4
Unclassified	1 068	10 158	5.8	699	8 152	6.4	605	6 985	4.5
Total	**5 631**	**175 526**	**100.0**	**3 631**	**126 421**	**100.0**	**3 832**	**153 979**	**100.0**

Northern Ireland County Statistics for Main Holdings
June Census 1991

LAND	Tyrone		
	holdings	hectares	area as % of N.Ireland
Total agricultural area	**7 009**	**245 662**	**24.3**
Total cereals	**577**	**4 228**	**9.2**
Wheat	27	205	3.5
Barley	534	3 737	10.1
Other	74	286	9.1
Total crops mainly for stockfeed	232	980	21.0
Other arable crops			
Potatoes	456	854	7.9
Oilseed rape	*	*	*
Horticultural crops			
Vegetables	32	116	8.8
Orchards and small fruit	66	91	4.8
HNS, bulbs and flowers	*	*	*
Glasshouse area	6	15	20.3
Grassland & all other land			
Grass <5 years old	4 400	43 042	24.2
All other grassland	6 682	186 702	25.4
All other land	2 251	9 570	32.6

LIVESTOCK			number as % of N.Ireland
	holdings	number	
Total cattle and calves	**6 373**	**360 428**	**23.5**
Dairy cows	1 674	63 078	23.0
Beef cows	4 379	63 749	25.1
Total pigs	**604**	**140 054**	**23.8**
Breeding herd	533	14 594	24.7
Total sheep and lambs	**2 695**	**583 649**	**22.7**
Breeding flock	2 580	271 116	22.5
Total fowls	**716**	**3 095 194**	**28.1**
Laying birds	530	619 892	22.6
Table birds	63	1 653 155	26.1

LABOUR			number as % of N.Ireland
	holdings	number	
Total labour force	**6 908**	**12 825**	**22.4**
Farmers, partners & directors	6 488	8 083	23.6
Regular whole-time workers	715	1 002	20.5
Regular part-time workers	915	1 208	22.5
Seasonal and casual workers	1 058	1 606	19.8

ANALYSIS BY TOTAL AREA	holdings	total area hectares	area as % of total
Under 20 ha	2 916	33 389	13.6
20 to <100 ha	3 796	160 876	65.5
100 to <300 ha	278	40 894	16.6
300 ha and over	19	10 504	4.3
Total	**7 009**	**245 663**	**100.0**

ANALYSIS BY FARM TYPE	holdings	total area hectares	area as % of total
Dairying	1 614	72 041	29.3
Beef cattle & sheep	3 619	147 165	59.9
Pigs and/or poultry	96	1 329	0.5
Mixed livestock	77	1 647	0.7
Crops and livestock	116	2 663	1.1
Cropping	102	3 553	1.4
Horticulture	58	1 306	0.5
Unclassified	1 327	15 958	6.5
Total	**7 009**	**245 662**	**100.0**

Chapter 6
Maps of selected main items; 1991

Note: based on main holdings only

Note on interpretation of maps

Ideally, geographical variation should be illustrated by small areas of approximately equal size in agricultural terms. However, in practice administrative boundaries must be used for convenience of collation. The following maps are based on such areas at county level. As counties are bound to vary somewhat in absolute size, all the maps are standardised by being based on comparative ratios of some kind, eg average size of holdings, wheat as a proportion of total agricultural area, sheep per 100 hectares of total area on holdings, and so on.

The maps use a simple shading system from light to dark which allows fine differences to be drawn. This is not the same as the traditional convention of map shading in which areas falling into a numerical size band are all given the same shade or hatching. The guide shadings on each map show convenient points along the scale. This system has the advantage that each map exhibits all gradations of difference without arbitrary jumps between bands; the eye can grasp more refinement of variation than in a size band system.

Key to County Maps

WESTERN ISLES

ORKNEY | SHETLAND

HIGHLAND

GRAMPIAN

TAYSIDE

FIFE

CENTRAL

LOTHIAN

STRATHCLYDE

BORDERS

DUMFRIES AND GALLOWAY

NORTHUMB

LONDON DERRY

ANTRIM

TYRONE

DOWN

TYNE AND WEAR

FERMANAGH

ARMAGH

CUMBRIA

DURHAM

CLEVELAND

NORTH YORKSHIRE

LANCASHIRE

WEST YORKS

HUMBERSIDE

GREATER MANCHESTER

SOUTH YORKS

MERSEYSIDE

CHESHIRE

DERBYS

NOTTS

LINCS

CLWYD

GWYNEDD

STAFFS

SHROPSHIRE

WEST MIDS

LEICS

NORFOLK

POWYS

HEREFORD AND WORCS

WARKS

NORTHANTS

CAMBS

SUFFOLK

BEDS

DYFED

GLOS

OXON

BUCKS

HERTS

WEST GLAM

MID GLAM

GWENT

ESSEX

SOUTH GLAM

AVON

BERKS

GREATER LONDON

WILTS

SURREY

KENT

SOMERSET

HANTS

WEST SUSSEX

EAST SUSSEX

DEVON

DORSET

ISLE OF WIGHT

CORNWALL & I.O.S.

Map 2

Average STANDARD GROSS MARGIN of holdings

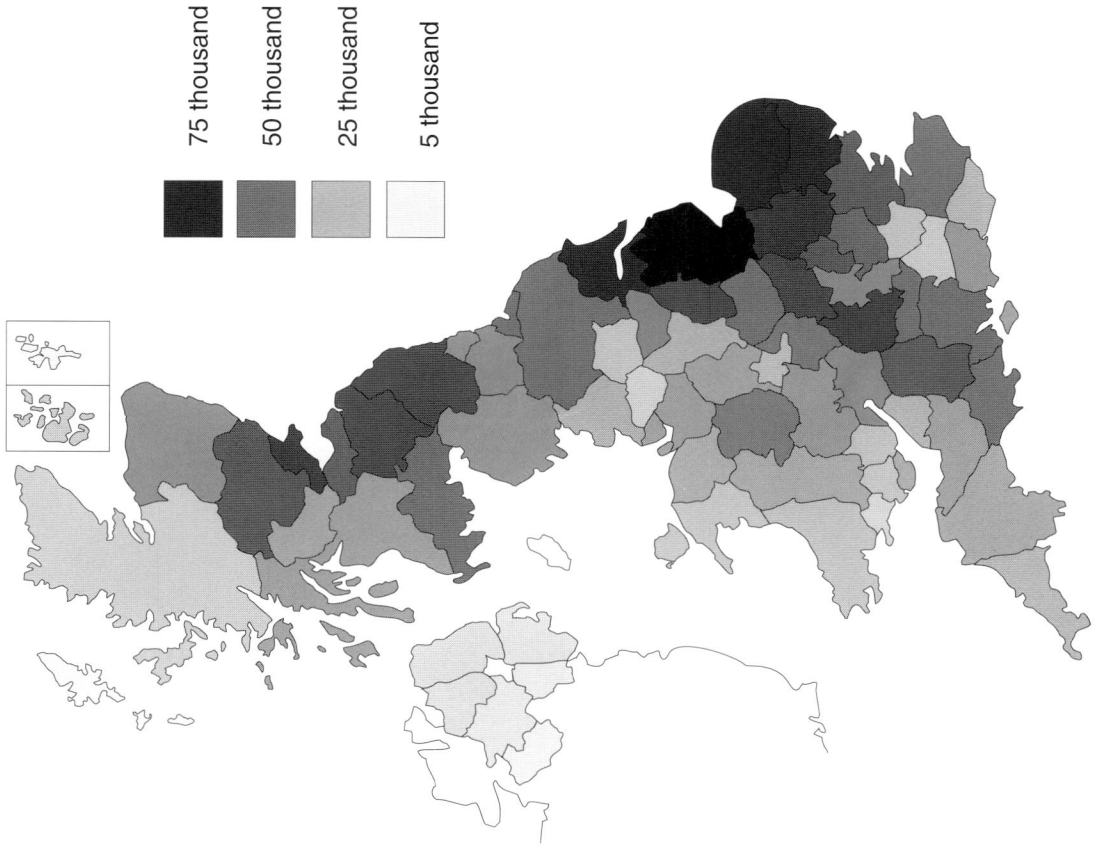

■	75 thousand
■	50 thousand
■	25 thousand
☐	5 thousand

Map 1

Average TOTAL AREA of holdings

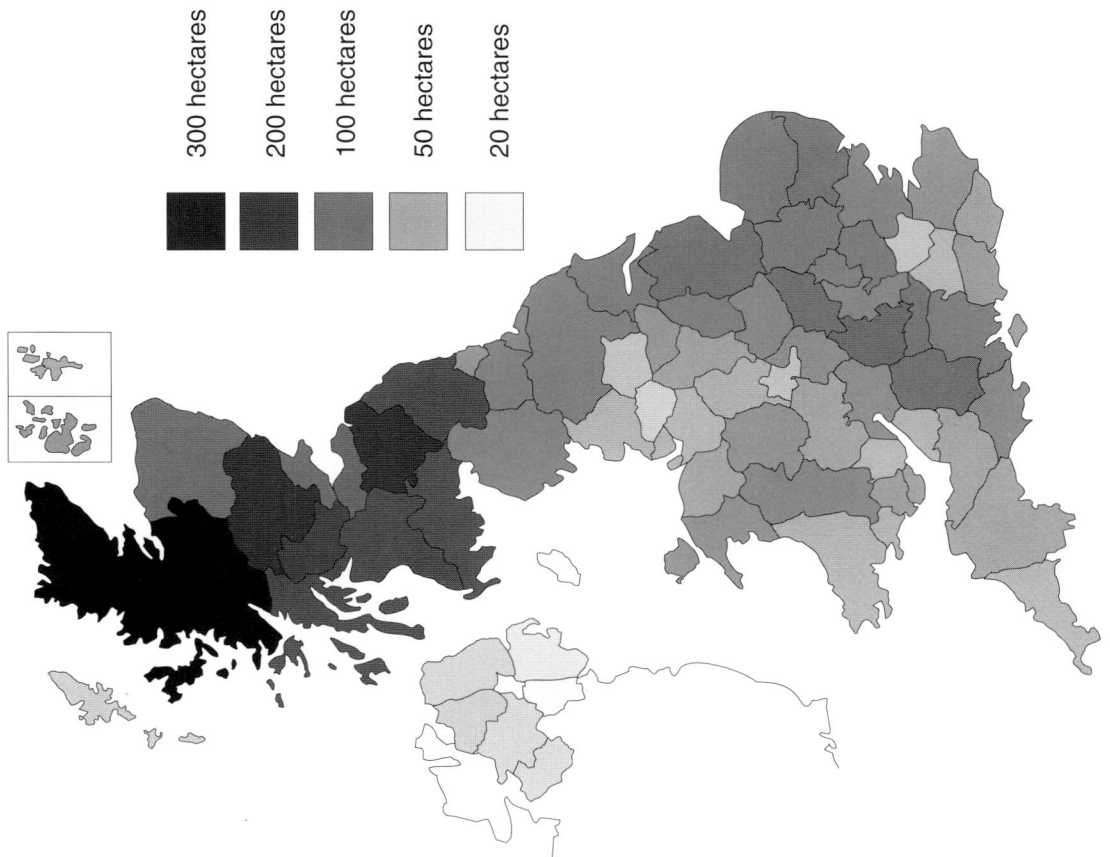

■	300 hectares
■	200 hectares
■	100 hectares
■	50 hectares
☐	20 hectares

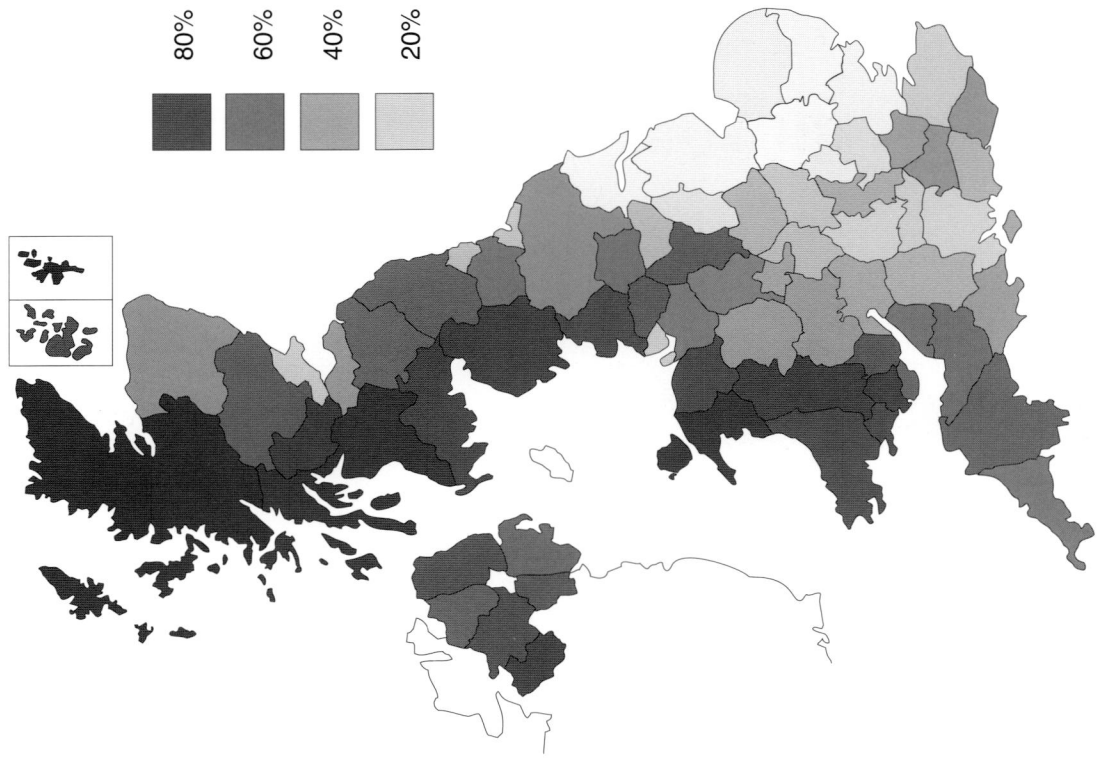

Map 4

Area of PERMANENT GRASSLAND
(as a percentage of total area)

80%
60%
40%
20%

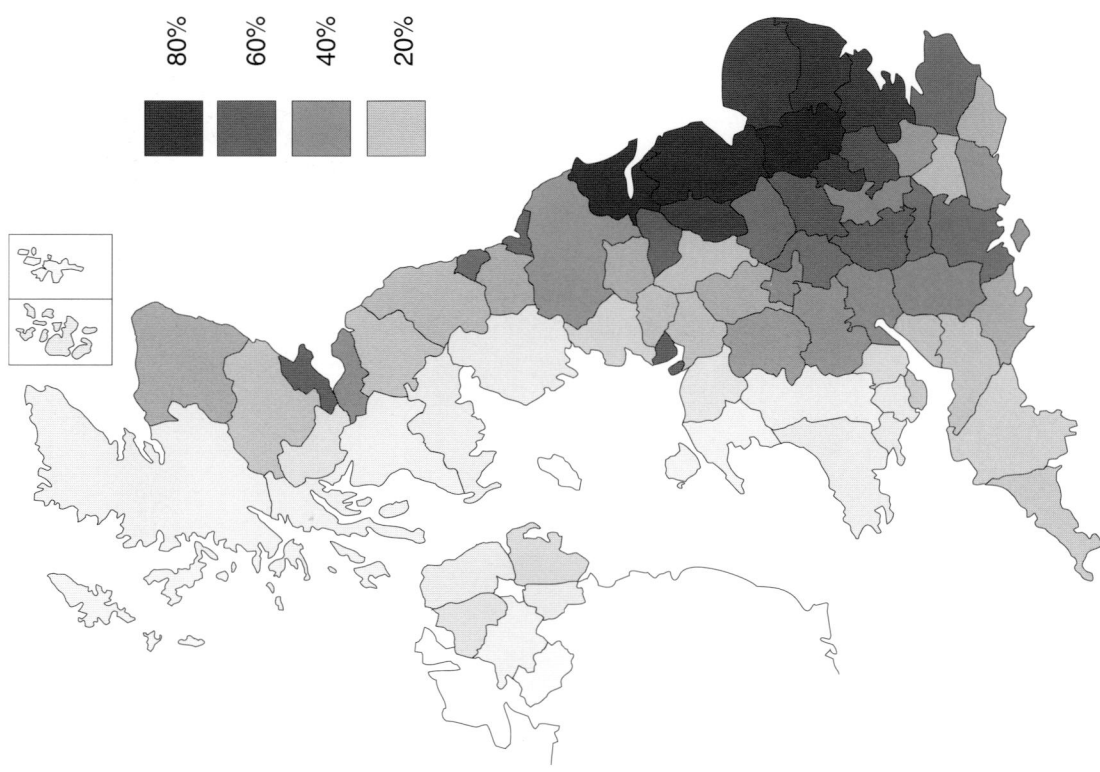

Map 3

Area of CROPS and FALLOW
(as a percentage of total area)

80%
60%
40%
20%

Map 6

Area of BARLEY
(as a percentage of total area)

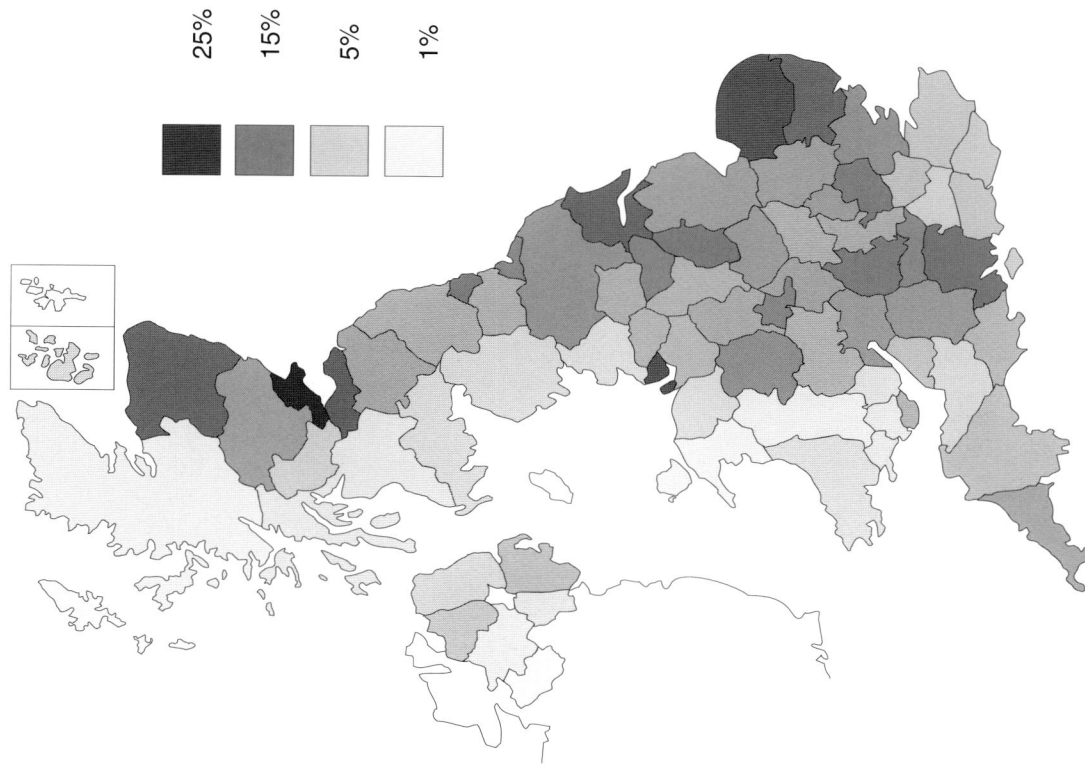

25%
15%
5%
1%

Map 5

Area of WHEAT
(as a percentage of total area)

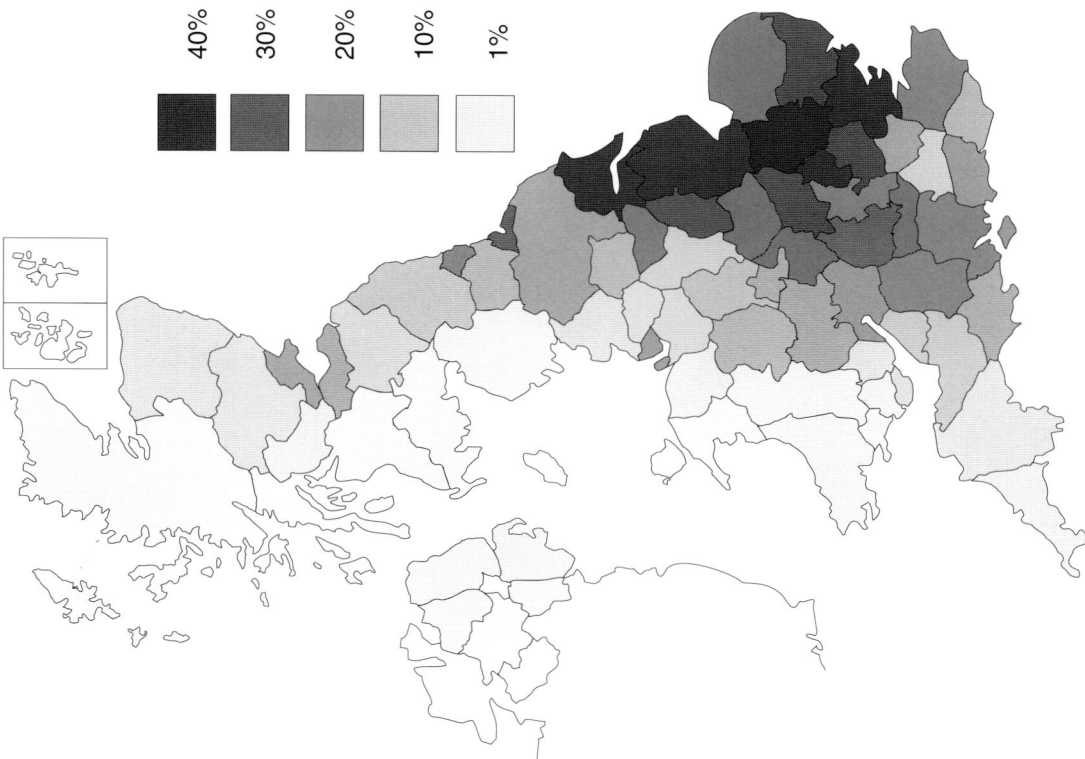

40%
30%
20%
10%
1%

Map 8

Area of FODDER CROPS
(as a percentage of total area)

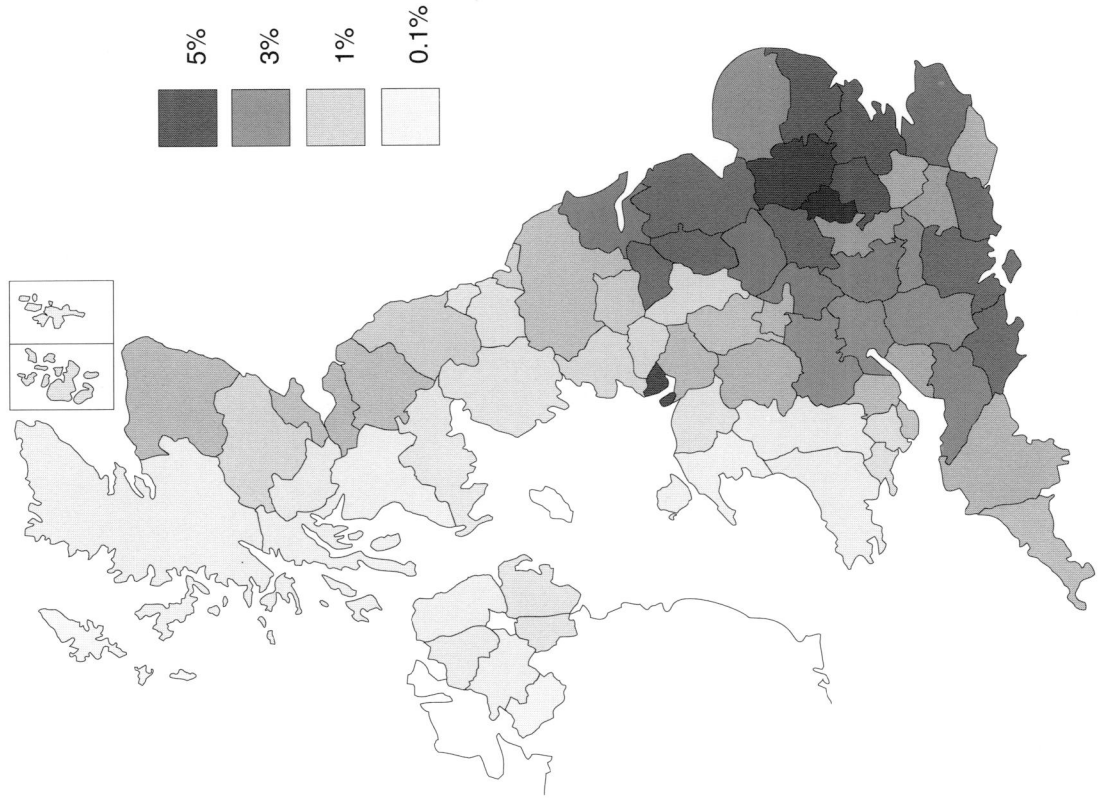

5%
3%
1%
0.1%

Map 7

Area of OILSEED RAPE
(as a percentage of total area)

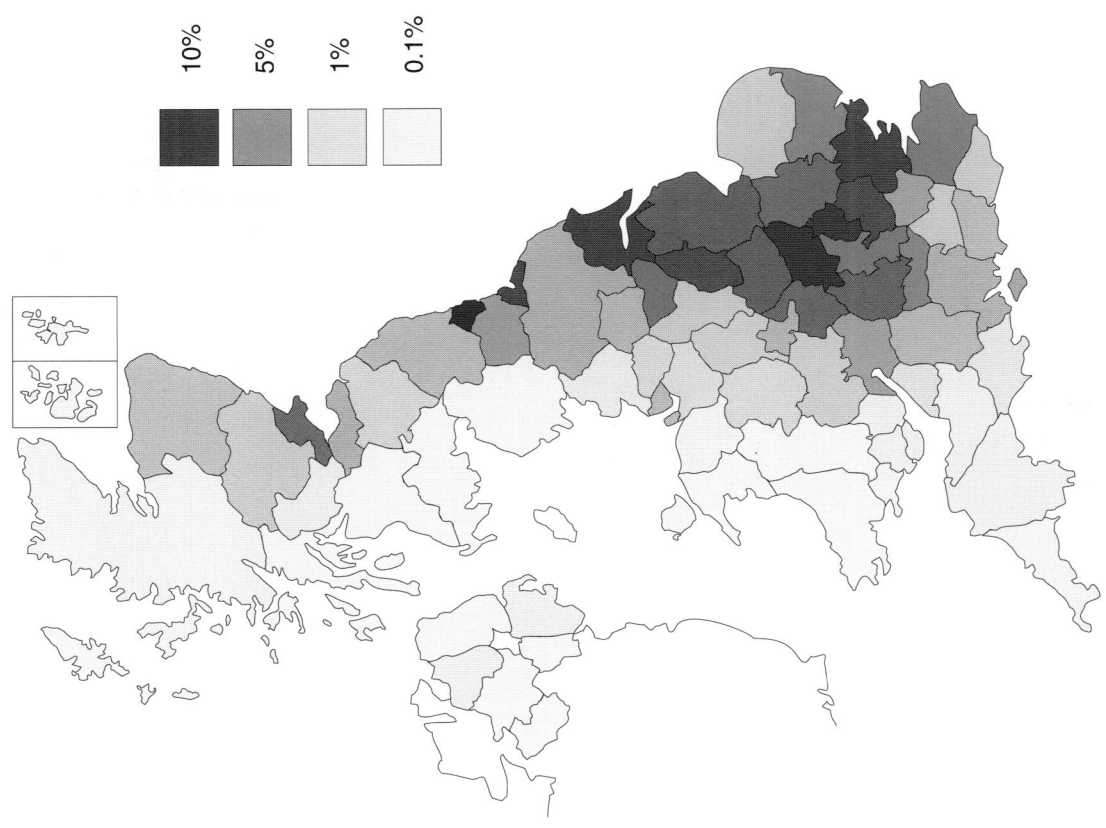

10%
5%
1%
0.1%

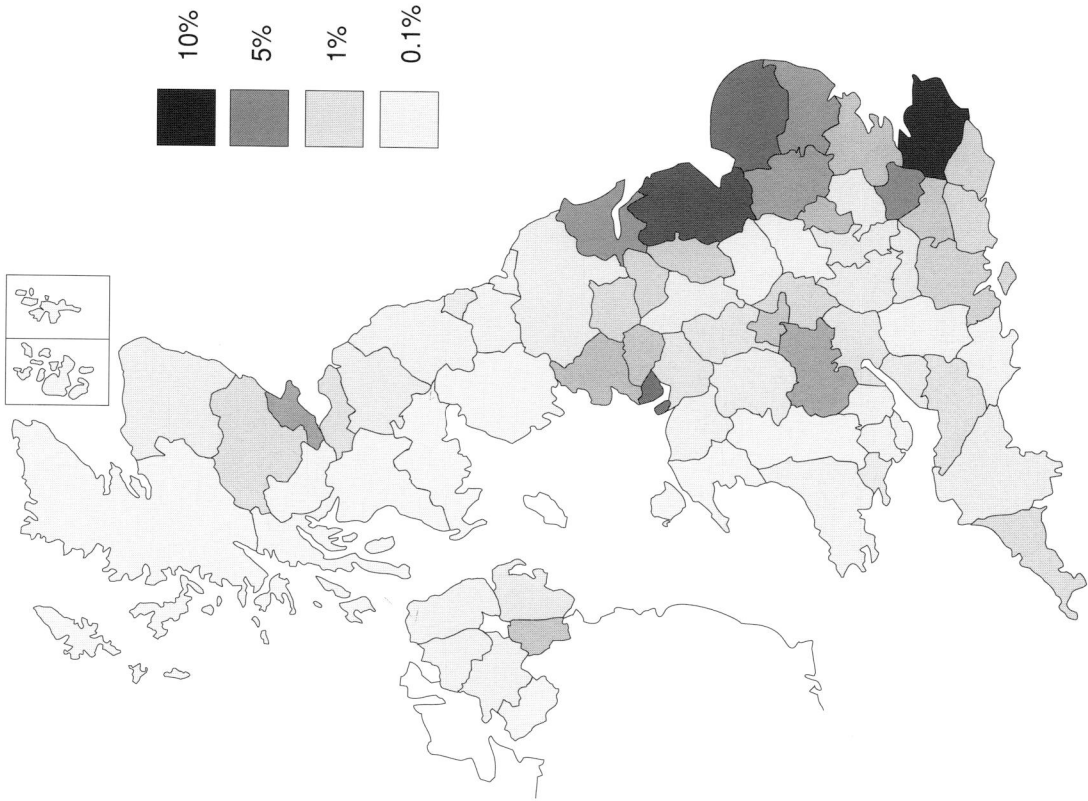

Map 10

Area of HORTICULTURAL CROPS
(as a percentage of total area)

10%
5%
1%
0.1%

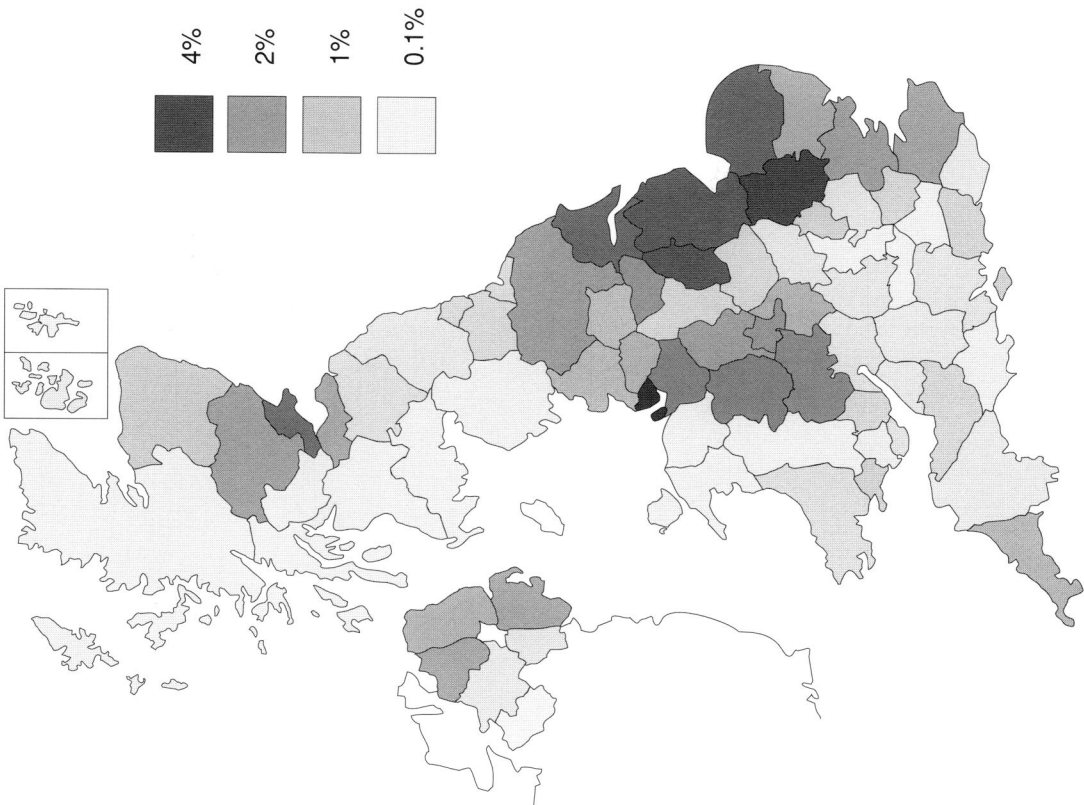

Map 9

Area of POTATOES
(as a percentage of total area)

4%
2%
1%
0.1%

Map 12

Number of BEEF COWS and HEIFERS
(per 100 hectares of total area)

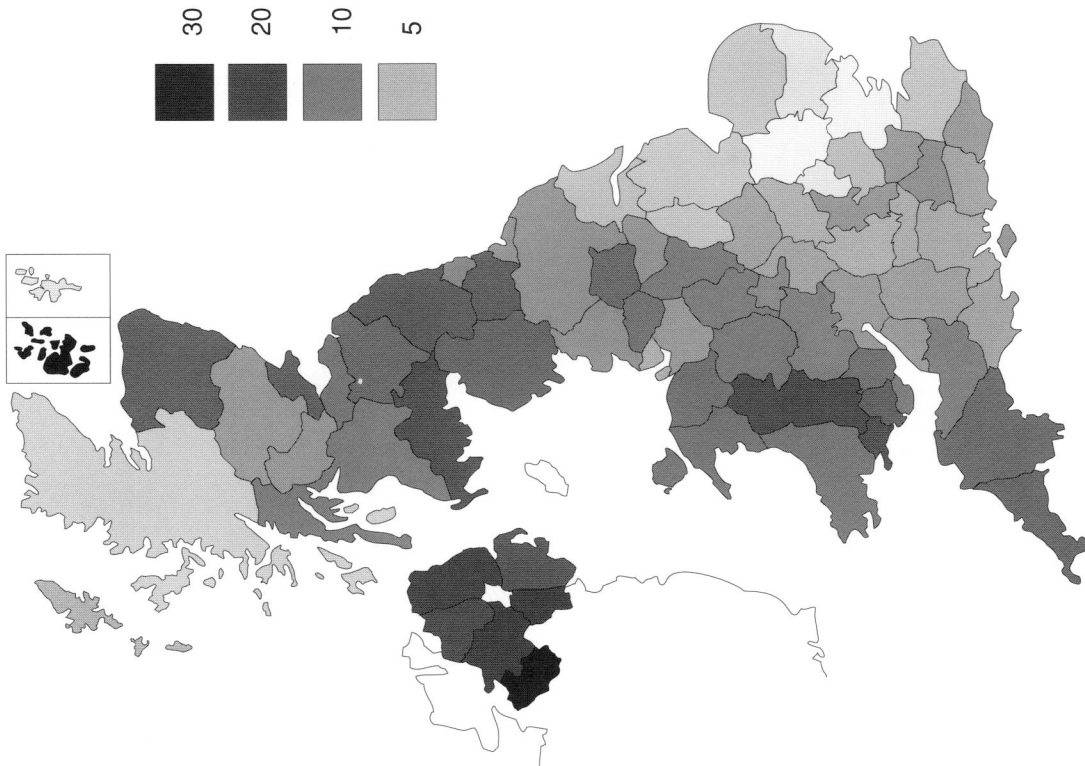

30
20
10
5

Map 11

Number of DAIRY COWS and HEIFERS
(per 100 hectares of total area)

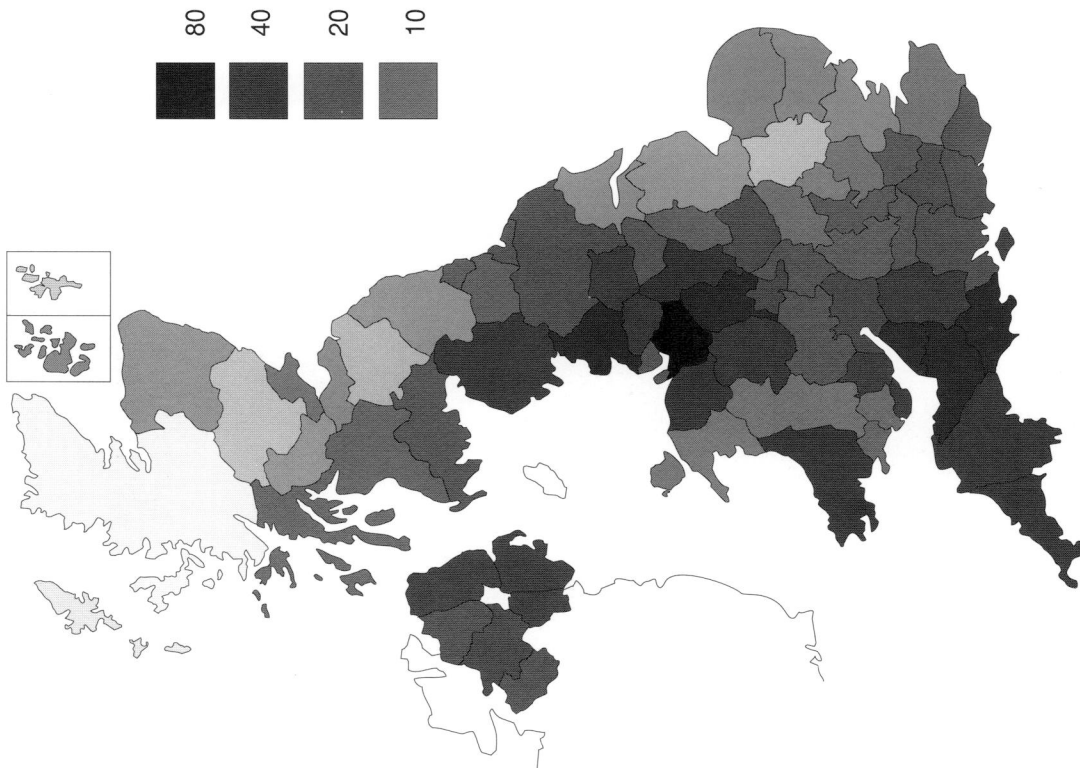

80
40
20
10

6-8

Map 14

Number of PIGS
(per 100 hectares of total area)

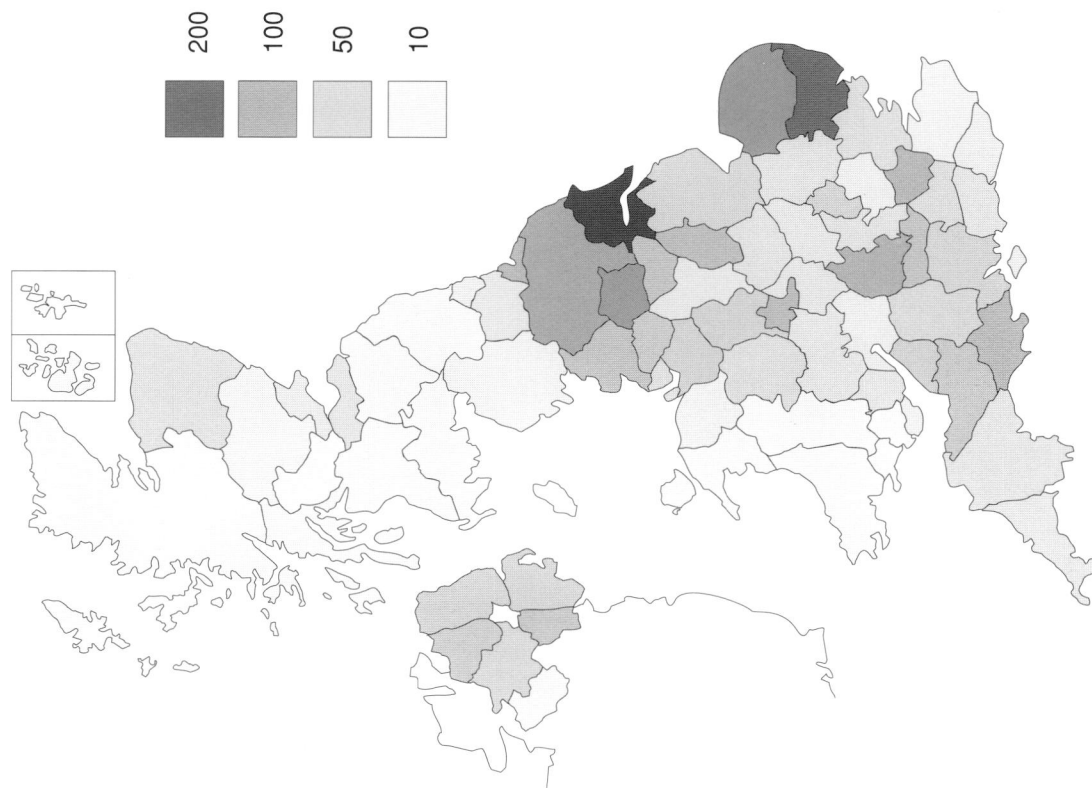

200
100
50
10

Map 13

Number of SHEEP
(per 100 hectares of total area)

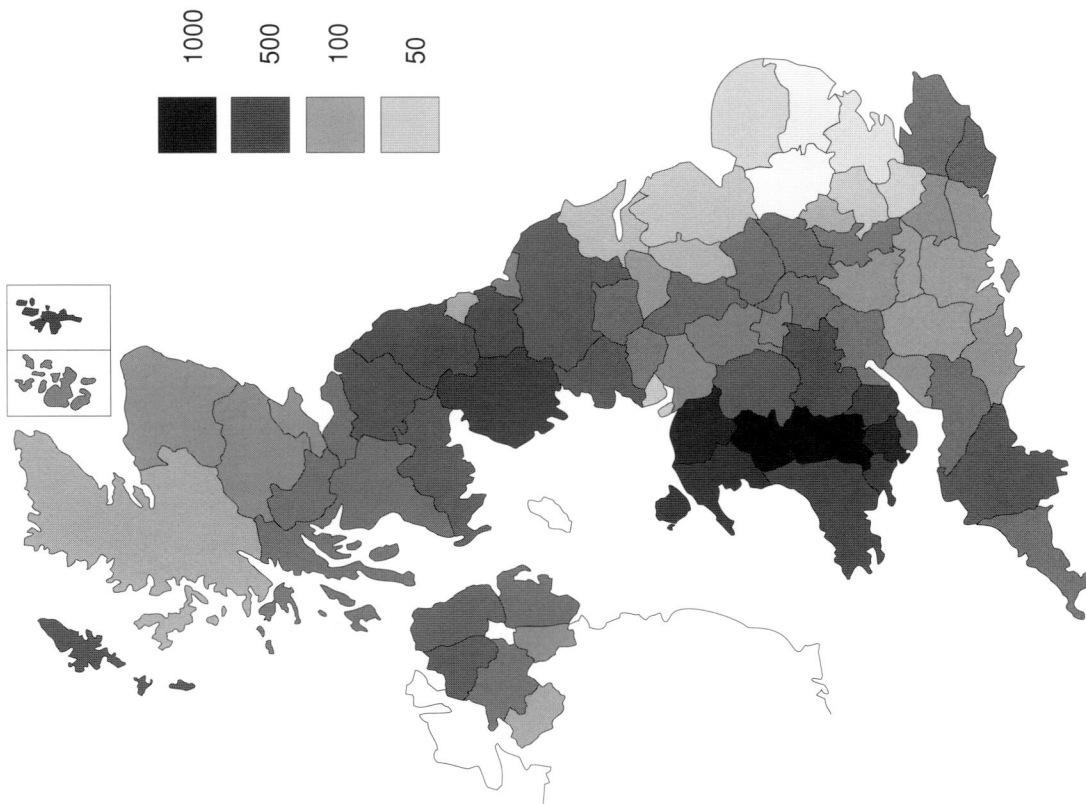

1000
500
100
50

Map 16

Number of REGULAR HIRED WORKERS
(as a percentage of total agricultural labour force)

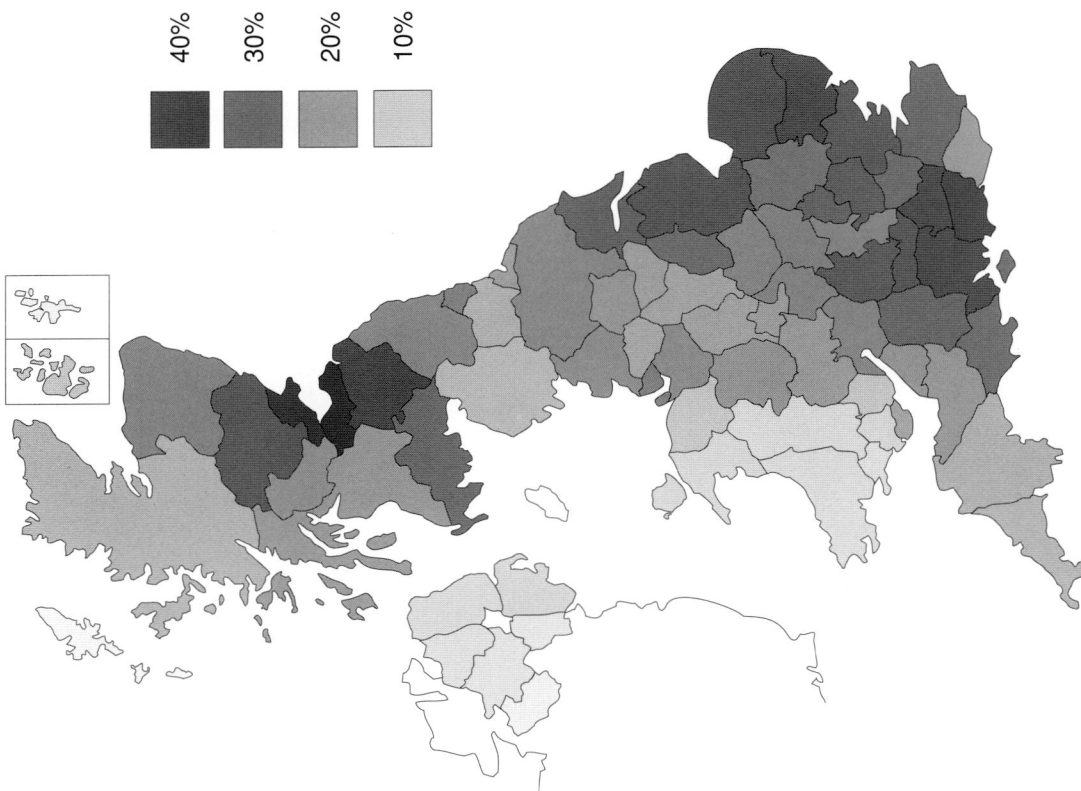

- 40%
- 30%
- 20%
- 10%

Map 15

Total AGRICULTURAL LABOUR FORCE
(per 1000 hectares of total area)

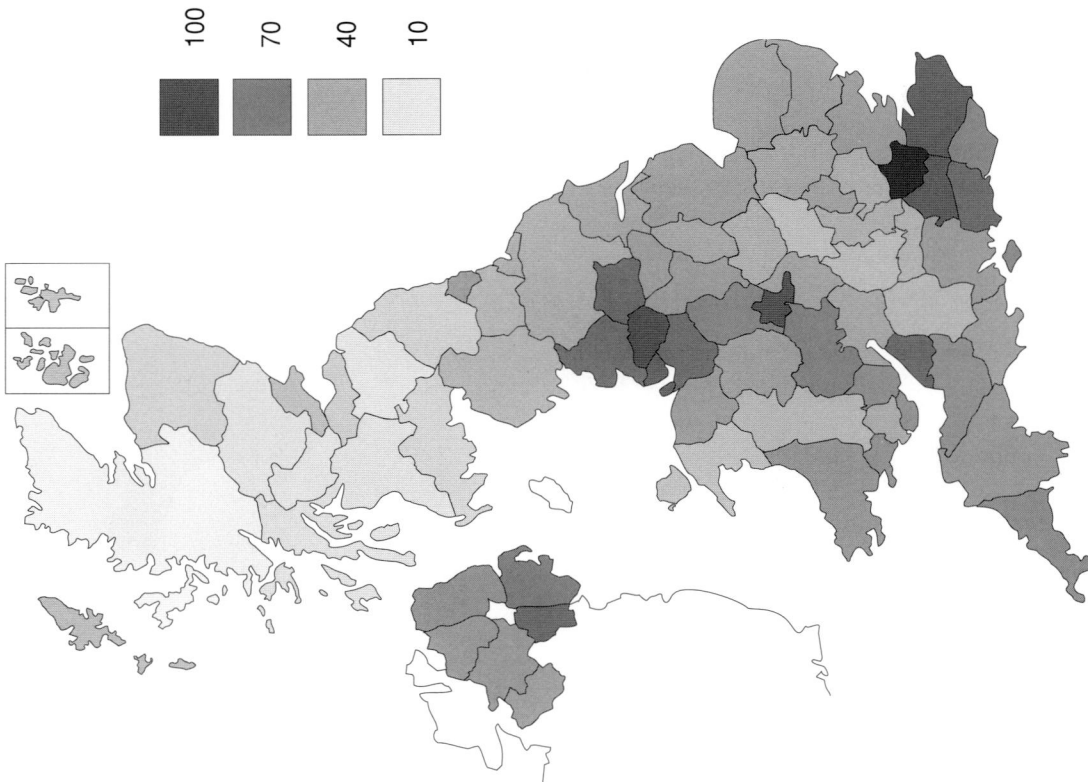

- 100
- 70
- 40
- 10

Chapter 7

Frequency distributions of main holdings by size; 1991

United Kingdom, England, Wales, Scotland, Northern Ireland

Table 7.1

Holdings by TOTAL AREA Size Groups
June Census 1991

	Under 2 hectares	2 - <5 hectares	5 - <10 hectares	10 - <20 hectares	20 - <30 hectares	30 - <40 hectares	40 - <50 hectares	50 - <100 hectares	100 - <200 hectares	200 - <300 hectares	300 - <500 hectares	500 - <700 hectares	700 and over hectares	Total
United Kingdom														
Number of holdings	13 121	19 439	28 643	36 739	25 319	19 244	15 797	42 452	25 411	7 123	4 420	1 344	1 889	240 941
percent	5.4%	8.1%	11.9%	15.2%	10.5%	8.0%	6.6%	17.6%	10.5%	3.0%	1.8%	0.6%	0.8%	100.0%
Area of holdings (ha.)	14 063	63 843	212 899	530 987	625 634	668 982	704 945	3 012 695	3 505 760	1 713 879	1 670 235	784 473	3 595 046	17 103 439
percent	0.1%	0.4%	1.2%	3.1%	3.7%	3.9%	4.1%	17.6%	20.5%	10.0%	9.8%	4.6%	21.0%	100.0%
England														
Number of holdings	9 917	13 787	18 255	21 400	14 405	11 294	9 525	26 298	16 773	4 869	2 952	829	662	150 966
percent	6.6%	9.1%	12.1%	14.2%	9.5%	7.5%	6.3%	17.4%	11.1%	3.2%	2.0%	0.5%	0.4%	100.0%
Area of holdings (ha.)	10 892	44 854	134 452	306 627	354 954	391 798	424 230	1 874 307	2 321 916	1 171 251	1 111 150	482 386	703 227	9 332 046
percent	0.1%	0.5%	1.4%	3.3%	3.8%	4.2%	4.5%	20.1%	24.9%	12.6%	11.9%	5.2%	7.5%	100.0%
Wales														
Number of holdings	951	2 002	3 988	4 877	3 474	2 698	2 283	5 828	2 680	516	296	65	52	29 710
percent	3.2%	6.7%	13.4%	16.4%	11.7%	9.1%	7.7%	19.6%	9.0%	1.7%	1.0%	0.2%	0.2%	100.0%
Area of holdings (ha.)	1 108	6 759	29 717	70 435	85 768	93 633	101 530	408 626	360 057	123 582	111 370	37 749	61 931	1 492 264
percent	0.1%	0.5%	2.0%	4.7%	5.7%	6.3%	6.8%	27.4%	24.1%	8.3%	7.5%	2.5%	4.2%	100.0%
Scotland														
Number of holdings	1 362	2 817	2 475	3 095	2 322	1 899	1 735	6 060	4 834	1 592	1 110	439	1 162	30 902
percent	4.4%	9.1%	8.0%	10.0%	7.5%	6.1%	5.6%	19.6%	15.6%	5.2%	3.6%	1.4%	3.8%	100.0%
Area of holdings (ha.)	1 666	9 256	17 789	44 844	57 487	66 044	77 373	437 698	673 552	383 930	424 430	257 861	2 817 647	5 269 574
percent	0.0%	0.2%	0.3%	0.9%	1.1%	1.3%	1.5%	8.3%	12.8%	7.3%	8.1%	4.9%	53.5%	100.0%
Northern Ireland														
Number of holdings	891	833	3 925	7 367	5 118	3 353	2 254	4 266	1 124	146	62	11	13	29 363
percent	3.0%	2.8%	13.4%	25.1%	17.4%	11.4%	7.7%	14.5%	3.8%	0.5%	0.2%	0.0%	0.0%	100.0%
Area of holdings (ha.)	397	2 974	30 941	109 081	127 425	117 507	101 812	292 064	150 235	35 116	23 285	6 477	12 241	1 009 555
percent	0.0%	0.3%	3.1%	10.8%	12.6%	11.6%	10.1%	28.9%	14.9%	3.5%	2.3%	0.6%	1.2%	100.0%

Note
Totals may not necessarily agree with the sum of their components due to rounding.

Table 7.2

Holdings by WHEAT AREA Size Groups
June Census 1991

	0.1 -<1 hectare	1 -<5 hectares	5 -<10 hectares	10 -<20 hectares	20 -<30 hectares	30 -<40 hectares	40 -<50 hectares	50 -<100 hectares	100 and over hectares	Total
United Kingdom										
Number of holdings	225	4 791	6 069	8 981	6 171	4 114	3 163	12 338		45 852
percent	0.5%	10.4%	13.2%	19.6%	13.5%	9.0%	6.9%	26.9%		100.0%
Area of wheat (ha.)	139	15 080	44 456	129 586	150 689	142 031	140 606	1 356 150		1 978 735
percent	0.0%	0.8%	2.2%	6.5%	7.6%	7.2%	7.1%	68.5%		100.0%
England										
Number of holdings	186	3 997	4 944	7 649	5 421	3 670	2 891	7 141	4 571	40 470
percent	0.5%	9.9%	12.2%	18.9%	13.4%	9.1%	7.1%	17.6%	11.3%	100.0%
Area of wheat (ha.)	118	12 502	36 236	110 719	132 424	126 784	128 511	503 650	800 450	1 851 393
percent	0.0%	0.7%	2.0%	6.0%	7.2%	6.8%	6.9%	27.2%	43.2%	100.0%
Wales										
Number of holdings	15	206	155	147	74	34	26	35	11	703
percent	2.1%	29.3%	22.0%	20.9%	10.5%	4.8%	3.7%	5.0%	1.6%	100.0%
Area of wheat (ha.)	7	631	1 079	2 041	1 806	1 171	1 163	2 447	1 436	11 780
percent	0.1%	5.4%	9.2%	17.3%	15.3%	9.9%	9.9%	20.8%	12.2%	100.0%
Scotland										
Number of holdings	9	333	818	1 088	629	393	239	449	122	4 080
percent	0.2%	8.2%	20.0%	26.7%	15.4%	9.6%	5.9%	11.0%	3.0%	100.0%
Area of wheat (ha.)	4	1 223	6 037	15 476	15 324	13 492	10 618	30 419	17 082	109 675
percent	0.0%	1.1%	5.5%	14.1%	14.0%	12.3%	9.7%	27.7%	15.6%	100.0%
Northern Ireland										
Number of holdings	15	255	152	97	47	17	7	9		599
percent	2.5%	42.6%	25.4%	16.2%	7.8%	2.8%	1.2%	1.5%		100.0%
Area of wheat (ha.)	10	724	1 104	1 350	1 135	584	314	666		5 887
percent	0.2%	12.3%	18.8%	22.9%	19.3%	9.9%	5.3%	11.3%		100.0%

Note
Totals may not necessarily agree with the sum of their components due to rounding.

Table 7.3

Holdings by BARLEY AREA Size Groups
June Census 1991

	0.1 - <1 hectare	1 - <5 hectares	5 - <10 hectares	10 - <20 hectares	20 - <30 hectares	30 - <40 hectares	40 - <50 hectares	50 - <100 hectares	100 and over hectares	Total
United Kingdom										
Number of holdings	686	12 159	11 655	14 745	8 154	4 659	2 988	6 556		61 602
percent	1.1%	19.7%	18.9%	23.9%	13.2%	7.6%	4.9%	10.6%		100.0%
Area of barley (ha.)	430	36 442	84 390	210 204	198 132	160 755	132 513	567 675		1 390 544
percent	0.0%	2.6%	6.1%	15.1%	14.2%	11.6%	9.5%	40.8%		100.0%
England										
Number of holdings	298	6 619	7 927	10 900	6 062	3 335	2 100	3 454	1 069	41 764
percent	0.7%	15.8%	19.0%	26.1%	14.5%	8.0%	5.0%	8.3%	2.6%	100.0%
Area of barley (ha.)	186	20 561	57 840	155 871	147 214	115 033	93 143	234 244	164 191	988 283
percent	0.0%	2.1%	5.9%	15.8%	14.9%	11.6%	9.4%	23.7%	16.6%	100.0%
Wales										
Number of holdings	48	1 350	899	688	256	112	40	53	9	3 455
percent	1.4%	39.1%	26.0%	19.9%	7.4%	3.2%	1.2%	1.5%	0.3%	100.0%
Area of barley (ha.)	30	4 022	6 332	9 446	6 160	3 797	1 741	3 465	1 149	36 143
percent	0.1%	11.1%	17.5%	26.1%	17.0%	10.5%	4.8%	9.6%	3.2%	100.0%
Scotland										
Number of holdings	95	1 640	1 693	2 466	1 621	1 132	816	1 523	396	11 382
percent	0.8%	14.4%	14.9%	21.7%	14.2%	9.9%	7.2%	13.4%	3.5%	100.0%
Area of barley (ha.)	52	5 150	12 390	35 393	39 591	39 184	36 220	103 332	57 801	329 115
percent	0.0%	1.6%	3.8%	10.8%	12.0%	11.9%	11.0%	31.4%	17.6%	100.0%
Northern Ireland										
Number of holdings	245	2 550	1 136	691	215	80	32	52		5 001
percent	4.9%	51.0%	22.7%	13.8%	4.3%	1.6%	0.6%	1.0%		100.0%
Area of barley (ha.)	162	6 709	7 828	9 494	5 167	2 741	1 409	3 493		37 003
percent	0.4%	18.1%	21.2%	25.7%	14.0%	7.4%	3.8%	9.4%		100.0%

Note

Totals may not necessarily agree with the sum of their components due to rounding.

Table 7.4

Holdings by TOTAL CEREALS AREA Size Groups
June Census 1991

	0.1 - <2 hectares	5 - <10 hectares	10 - <20 hectares	20 - <30 hectares	30 - <40 hectares	40 - <50 hectares	50 - <100 hectares	100 - <150 hectares	150 and over hectares	Total
United Kingdom										
Number of holdings	3 841	10 543	13 154	8 408	6 028	4 745	12 297		9 313	78 074
percent	4.9%	13.5%	16.8%	10.8%	7.7%	6.1%	15.8%		11.9%	100.0%
Area of cereals (ha.)	4 313	75 839	188 951	206 479	208 569	211 264	870 905		1 697 000	3 495 872
percent	0.1%	2.2%	5.4%	5.9%	6.0%	6.0%	24.9%		48.5%	100.0%
England										
Number of holdings	1 275	6 651	9 242	6 400	4 742	3 755	10 051	4 065	4 159	55 369
percent	2.3%	12.0%	16.7%	11.6%	8.6%	6.8%	18.2%	7.3%	7.5%	100.0%
Area of cereals (ha.)	1 578	48 190	133 549	157 427	164 273	167 192	713 227	494 659	1 032 166	2 929 344
percent	0.1%	1.6%	4.6%	5.4%	5.6%	5.7%	24.3%	16.9%	35.2%	100.0%
Wales										
Number of holdings	349	954	770	280	163	91	148	19	14	3 951
percent	8.8%	24.1%	19.5%	7.1%	4.1%	2.3%	3.7%	0.5%	0.4%	100.0%
Area of cereals (ha.)	446	6 686	10 706	6 731	5 548	3 991	10 039	2 306	2 722	53 029
percent	0.8%	12.6%	20.2%	12.7%	10.5%	7.5%	18.9%	4.3%	5.1%	100.0%
Scotland										
Number of holdings	1 028	1 757	2 390	1 482	1 021	834	2 005	657	386	13 254
percent	7.8%	13.3%	18.0%	11.2%	7.7%	6.3%	15.1%	5.0%	2.9%	100.0%
Area of cereals (ha.)	897	12 787	34 193	36 343	35 254	37 162	141 511	79 126	84 505	467 471
percent	0.2%	2.7%	7.3%	7.8%	7.5%	7.9%	30.3%	16.9%	18.1%	100.0%
Northern Ireland										
Number of holdings	1 189	1 181	752	246	102	65	93		13	5 500
percent	21.6%	21.5%	13.7%	4.5%	1.9%	1.2%	1.7%		0.2%	100.0%
Area of cereals (ha.)	1 392	8 176	10 503	5 978	3 494	2 919	6 128		1 516	46 028
percent	3.0%	17.8%	22.8%	13.0%	7.6%	6.3%	13.3%		3.3%	100.0%

Notes

Maize is excluded from this table.

Totals may not necessarily agree with the sum of their components due to rounding.

Table 7.5

Holdings by POTATO AREA Size Groups
June Census 1991

	0.1 - <1 hectare	1 - <5 hectares	5 - <10 hectares	10 - <20 hectares	20 - <40 hectares	40 and over hectares	Total
United Kingdom							
Number of holdings	9 021	7 708	4 126	3 092	1 609	625	26 181
percent	34.5%	29.4%	15.8%	11.8%	6.1%	2.4%	100.0%
Area of potatoes (ha.)	2 872	20 136	29 364	42 538	43 534	37 829	176 272
percent	1.6%	11.4%	16.7%	24.1%	24.7%	21.5%	100.0%
England							
Number of holdings	3 924	5 340	2 855	2 246	1 282	540	16 187
percent	24.2%	33.0%	17.6%	13.9%	7.9%	3.3%	100.0%
Area of potatoes (ha.)	1 500	14 057	20 242	30 944	34 930	33 293	134 964
percent	1.1%	10.4%	15.0%	22.9%	25.9%	24.7%	100.0%
Wales							
Number of holdings	1 015	394	112	76	31	6	1 634
percent	62.1%	24.1%	6.9%	4.7%	1.9%	0.4%	100.0%
Area of potatoes (ha.)	306	939	783	1 033	823	295	4 179
percent	7.3%	22.5%	18.7%	24.7%	19.7%	7.1%	100.0%
Scotland							
Number of holdings	3 002	879	829	599	233	55	5 597
percent	53.6%	15.7%	14.8%	10.7%	4.2%	1.0%	100.0%
Area of potatoes (ha.)	645	2 535	5 985	8 186	6 094	2 925	26 370
percent	2.4%	9.6%	22.7%	31.0%	23.1%	11.1%	100.0%
Northern Ireland							
Number of holdings	1 080	1 095	330	171	63	24	2 763
percent	39.1%	39.6%	11.9%	6.2%	2.3%	0.9%	100.0%
Area of potatoes (ha.)	421	2 605	2 354	2 375	1 687	1 316	10 759
percent	3.9%	24.2%	21.9%	22.1%	15.7%	12.2%	100.0%

Note
Totals may not necessarily agree with the sum of their components due to rounding.

Holdings by OILSEED RAPE AREA Size Groups
June Census 1991

Table 7.6

	0.1 - <5 hectares	5 - <10 hectares	10 - <20 hectares	20 - <50 hectares	50 - <100 hectares	100 and over hectares	Total
United Kingdom							
Number of holdings	819	2 652	5 079	7 546		377	16 473
percent	5.0%	16.1%	30.8%	45.7%		2.3%	100.0%
Area of oilseed rape (ha.)	2 787	20 158	72 985	289 813		53 883	439 625
percent	0.6%	4.6%	16.6%	65.9%		12.3%	100.0%
England							
Number of holdings	609	2 160	4 263	5 143	1 479	352	14 006
percent	4.3%	15.4%	30.4%	36.7%	10.6%	2.5%	100.0%
Area of oilseed rape (ha.)	2 106	16 469	61 404	158 227	98 674	50 615	387 495
percent	0.5%	4.3%	15.8%	40.8%	25.5%	13.1%	100.0%
Wales							
Number of holdings	12	8	24	13	5		62
percent	19.4%	12.9%	38.7%	21.0%	8.1%	0.0%	100.0%
Area of oilseed rape (ha.)	26	59	332	377	290		1 084
percent	2.4%	5.4%	30.6%	34.8%	26.8%	0.0%	100.0%
Scotland							
Number of holdings	149	445	765	764	131	25	2 279
percent	6.5%	19.5%	33.6%	33.5%	5.7%	1.1%	100.0%
Area of oilseed rape (ha.)	524	3 369	10 892	23 127	8 717	3 268	49 895
percent	1.1%	6.8%	21.8%	46.4%	17.5%	6.5%	100.0%
Northern Ireland							
Number of holdings	49	39	27	11			126
percent	38.9%	31.0%	21.4%	8.7%		0.0%	100.0%
Area of oilseed rape (ha.)	131	261	357	401			1 151
percent	11.4%	22.7%	31.0%	34.9%		0.0%	100.0%

Note

Totals may not necessarily agree with the sum of their components due to rounding.

Table 7.7

Holdings by TOTAL HORTICULTURE AREA Size Groups
June Census 1991

	0.1 - <1 hectare	1 - <2 hectares	2 - <5 hectares	5 - <20 hectares	20 - <50 hectares	50 and over hectares	Total
United Kingdom							
Number of holdings	7 977	3 285	4 280	4 952	1 879	737	23 110
percent	34.5%	14.2%	18.5%	21.4%	8.1%	3.2%	100.0%
Area of hort. crops (ha.)	3 071	4 401	13 289	51 677	57 674	71 952	202 065
percent	1.5%	2.2%	6.6%	25.6%	28.5%	35.6%	100.0%
England							
Number of holdings	6 539	2 799	3 637	4 224	1 673	695	19 567
percent	33.4%	14.3%	18.6%	21.6%	8.6%	3.6%	100.0%
Area of hort. crops (ha.)	2 590	3 748	11 280	44 271	51 385	68 540	181 814
percent	1.4%	2.1%	6.2%	24.3%	28.3%	37.7%	100.0%
Wales							
Number of holdings	326	93	84	66	9	3	581
percent	56.1%	16.0%	14.5%	11.4%	1.5%	0.5%	100.0%
Area of hort. crops (ha.)	118	122	245	531	276	217	1 510
percent	7.8%	8.1%	16.2%	35.2%	18.3%	14.4%	100.0%
Scotland							
Number of holdings	748	155	277	484	183	39	1 886
percent	39.7%	8.2%	14.7%	25.7%	9.7%	2.1%	100.0%
Area of hort. crops (ha.)	209	208	896	5 250	5 572	3 195	15 330
percent	1.4%	1.4%	5.8%	34.2%	36.3%	20.8%	100.0%
Northern Ireland							
Number of holdings	364	238	282	178	14		1 076
percent	33.8%	22.1%	26.2%	16.5%	1.3%	0.0%	100.0%
Area of hort. crops (ha.)	154	323	868	1 625	441		3 411
percent	4.5%	9.5%	25.4%	47.6%	12.9%	0.0%	100.0%

Notes

Mushrooms are excluded from this table.

Totals may not necessarily agree with the sum of their components due to rounding.

Table 7.8

Holdings by Size of DAIRY BREEDING HERD
June Census 1991

	1 - <10	10 - <30	30 - <40	40 - <50	50 - <70	70 - <100	100 - <200	200 and over	Total
United Kingdom									
Number of holdings	3 434	7 708	4 980	4 860	7 638	7 564	6 779	943	43 906
percent	7.8%	17.6%	11.3%	11.1%	17.4%	17.2%	15.4%	2.1%	100.0%
Number of cows	12 936	154 251	171 071	213 675	446 965	626 016	882 069	261 150	2 768 133
percent	0.5%	5.6%	6.2%	7.7%	16.1%	22.6%	31.9%	9.4%	100.0%
England									
Number of holdings	1 899	3 880	2 885	2 963	5 202	5 340	5 027	767	27 963
percent	6.8%	13.9%	10.3%	10.6%	18.6%	19.1%	18.0%	2.7%	100.0%
Number of cows	6 486	79 491	99 397	130 565	305 099	443 107	658 097	214 074	1 936 316
percent	0.3%	4.1%	5.1%	6.7%	15.8%	22.9%	34.0%	11.1%	100.0%
Wales									
Number of holdings	474	1 490	888	807	1 016	821	583	62	6 141
percent	7.7%	24.3%	14.5%	13.1%	16.5%	13.4%	9.5%	1.0%	100.0%
Number of cows	1 935	30 283	30 644	35 562	59 108	66 979	75 230	17 350	317 091
percent	0.6%	9.6%	9.7%	11.2%	18.6%	21.1%	23.7%	5.5%	100.0%
Scotland									
Number of holdings	435	144	150	255	492	735	818	89	3 118
percent	14.0%	4.6%	4.8%	8.2%	15.8%	23.6%	26.2%	2.9%	100.0%
Number of cows	1 046	2 925	5 186	11 357	29 178	61 599	105 405	23 972	240 668
percent	0.4%	1.2%	2.2%	4.7%	12.1%	25.6%	43.8%	10.0%	100.0%
Northern Ireland									
Number of holdings	626	2 194	1 057	835	928	668	351	25	6 684
percent	9.4%	32.8%	15.8%	12.5%	13.9%	10.0%	5.3%	0.4%	100.0%
Number of cows	3 469	41 552	35 844	36 191	53 580	54 331	43 337	5 754	274 058
percent	1.3%	15.2%	13.1%	13.2%	19.6%	19.8%	15.8%	2.1%	100.0%

Table 7.9

Holdings by Size of BEEF BREEDING HERD
June Census 1991

	1 - <10	10 - <30	30 - <40	40 - <50	50 - <100	100 and over	Total
United Kingdom							
Number of holdings	33 398	23 746	5 216	3 512	6 514	2 294	74 680
percent	44.7%	31.8%	7.0%	4.7%	8.7%	3.1%	100.0%
Number of cows	137 261	411 050	175 795	153 247	439 098	338 578	1 655 029
percent	8.3%	24.8%	10.6%	9.3%	26.5%	20.5%	100.0%
England							
Number of holdings	16 959	10 375	2 300	1 567	2 821	865	34 887
percent	48.6%	29.7%	6.6%	4.5%	8.1%	2.5%	100.0%
Number of cows	65 580	179 660	77 545	68 559	189 093	127 766	708 203
percent	9.3%	25.4%	10.9%	9.7%	26.7%	18.0%	100.0%
Wales							
Number of holdings	5 077	3 996	874	534	724	110	11 315
percent	44.9%	35.3%	7.7%	4.7%	6.4%	1.0%	100.0%
Number of cows	21 040	70 539	29 378	23 203	46 859	15 177	206 196
percent	10.2%	34.2%	14.2%	11.3%	22.7%	7.4%	100.0%
Scotland							
Number of holdings	3 332	3 074	1 033	892	2 351	1 194	11 876
percent	28.1%	25.9%	8.7%	7.5%	19.8%	10.1%	100.0%
Number of cows	13 846	55 457	35 034	39 053	164 230	178 720	486 340
percent	2.8%	11.4%	7.2%	8.0%	33.8%	36.7%	100.0%
Northern Ireland							
Number of holdings	8 030	6 301	1 009	519	618	125	16 602
percent	48.4%	38.0%	6.1%	3.1%	3.7%	0.8%	100.0%
Number of cows	36 795	105 394	33 838	22 432	38 916	16 915	254 290
percent	14.5%	41.4%	13.3%	8.8%	15.3%	6.7%	100.0%

Table 7.10

Holdings by Total Number of CATTLE and CALVES
June Census 1991

	1 - <10	10 - <30	30 - <40	40 - <50	50 - <70	70 - <100	100 - <200	200 and over	Total
United Kingdom									
Number of holdings	16 423	29 823	10 952	9 434	15 352	17 164	27 451	13 721	140 320
percent	11.7%	21.3%	7.8%	6.7%	10.9%	12.2%	19.6%	9.8%	100.0%
Number of cattle	84 569	554 379	374 443	416 817	904 925	1 433 870	3 838 266	4 200 626	11 807 895
percent	0.7%	4.7%	3.2%	3.5%	7.7%	12.1%	32.5%	35.6%	100.0%
England									
Number of holdings	9 498	15 336	5 612	4 880	8 179	9 600	16 580	8 131	77 816
percent	12.2%	19.7%	7.2%	6.3%	10.5%	12.3%	21.3%	10.4%	100.0%
Number of cattle	47 583	284 271	191 929	215 616	482 657	802 518	2 324 623	2 482 330	6 831 527
percent	0.7%	4.2%	2.8%	3.2%	7.1%	11.7%	34.0%	36.3%	100.0%
Wales									
Number of holdings	2 403	4 238	1 620	1 500	2 398	2 644	3 378	1 058	19 239
percent	12.5%	22.0%	8.4%	7.8%	12.5%	13.7%	17.6%	5.5%	100.0%
Number of cattle	12 507	78 692	55 606	66 490	141 470	220 604	460 078	299 829	1 335 276
percent	0.9%	5.9%	4.2%	5.0%	10.6%	16.5%	34.5%	22.5%	100.0%
Scotland									
Number of holdings	1 817	2 800	1 010	849	1 588	2 004	4 102	3 485	17 655
percent	10.3%	15.9%	5.7%	4.8%	9.0%	11.4%	23.2%	19.7%	100.0%
Number of cattle	9 050	51 474	34 346	37 440	93 825	168 140	591 443	1 122 148	2 107 866
percent	0.4%	2.4%	1.6%	1.8%	4.5%	8.0%	28.1%	53.2%	100.0%
Northern Ireland									
Number of holdings	2 705	7 449	2 710	2 205	3 187	2 916	3 391	1 047	25 610
percent	10.6%	29.1%	10.6%	8.6%	12.4%	11.4%	13.2%	4.1%	100.0%
Number of cattle	15 429	139 942	92 562	97 271	186 973	242 608	462 122	296 319	1 533 226
percent	1.0%	9.1%	6.0%	6.3%	12.2%	15.8%	30.1%	19.3%	100.0%

Table 7.11

Holdings by Size of PIG BREEDING HERD
June Census 1991

	1 - <10	10 - <20	20 - <50	50 - <100	100 - <200	200 and over	Total
United Kingdom							
Number of holdings	5 811	1 346	1 665	1 314	1 263	1 030	12 429
percent	46.8%	10.8%	13.4%	10.6%	10.2%	8.3%	100.0%
Number of pigs	19 301	18 227	53 364	94 206	175 239	422 592	782 929
percent	2.5%	2.3%	6.8%	12.0%	22.4%	54.0%	100.0%
England							
Number of holdings	3 500	850	1 223	1 023	1 098	906	8 600
percent	40.7%	9.9%	14.2%	11.9%	12.8%	10.5%	100.0%
Number of pigs	11 725	11 710	39 736	74 362	152 738	370 422	660 693
percent	1.8%	1.8%	6.0%	11.3%	23.1%	56.1%	100.0%
Wales							
Number of holdings	763	81	54	27	17	8	950
percent	80.3%	8.5%	5.7%	2.8%	1.8%	0.8%	100.0%
Number of pigs	2 032	1 041	1 755	1 825	2 309	2 787	11 749
percent	17.3%	8.9%	14.9%	15.5%	19.7%	23.7%	100.0%
Scotland							
Number of holdings	225	50	64	77	60	70	546
percent	41.2%	9.2%	11.7%	14.1%	11.0%	12.8%	100.0%
Number of pigs	708	673	1 969	5 319	8 413	34 293	51 375
percent	1.4%	1.3%	3.8%	10.4%	16.4%	66.8%	100.0%
Northern Ireland							
Number of holdings	1 323	365	324	187	88	46	2 333
percent	56.7%	15.6%	13.9%	8.0%	3.8%	2.0%	100.0%
Number of pigs	4 836	4 803	9 904	12 700	11 779	15 090	59 112
percent	8.2%	8.1%	16.8%	21.5%	19.9%	25.5%	100.0%

Table 7.12

Holdings by Total Number of PIGS
June Census 1991

	1 - <10	10 - <20	20 - <50	50 - <100	100 - <200	200 - <500	500 - <1000	1000 and over	Total
United Kingdom									
Number of holdings	4 738	1 736	2 144	1 421	1 449	2 032	1 539	2 167	17 226
percent	27.5%	10.1%	12.4%	8.2%	8.4%	11.8%	8.9%	12.6%	100.0%
Number of pigs	16 174	23 471	67 965	101 004	208 159	665 889	1 101 670	5 394 179	7 578 511
percent	0.2%	0.3%	0.9%	1.3%	2.7%	8.8%	14.5%	71.2%	100.0%
England									
Number of holdings	3 080	1 148	1 394	916	1 083	1 573	1 305	1 864	12 363
percent	24.9%	9.3%	11.3%	7.4%	8.8%	12.7%	10.6%	15.1%	100.0%
Number of pigs	11 130	15 560	44 565	65 710	156 196	519 189	940 284	4 642 465	6 395 099
percent	0.2%	0.2%	0.7%	1.0%	2.4%	8.1%	14.7%	72.6%	100.0%
Wales									
Number of holdings	769	205	174	83	49	53	24	18	1 375
percent	55.9%	14.9%	12.7%	6.0%	3.6%	3.9%	1.7%	1.3%	100.0%
Number of pigs	2 345	2 691	5 370	5 696	6 737	17 035	16 501	45 663	102 038
percent	2.3%	2.6%	5.3%	5.6%	6.6%	16.7%	16.2%	44.8%	100.0%
Scotland									
Number of holdings	303	77	68	65	69	112	76	132	902
percent	33.6%	8.5%	7.5%	7.2%	7.6%	12.4%	8.4%	14.6%	100.0%
Number of pigs	935	1 035	2 114	4 565	10 104	37 528	53 009	383 735	493 025
percent	0.2%	0.2%	0.4%	0.9%	2.0%	7.6%	10.8%	77.8%	100.0%
Northern Ireland									
Number of holdings	586	306	508	357	248	294	134	153	2 586
percent	22.7%	11.8%	19.6%	13.8%	9.6%	11.4%	5.2%	5.9%	100.0%
Number of pigs	1 764	4 185	15 916	25 033	35 122	92 137	91 876	322 316	588 349
percent	0.3%	0.7%	2.7%	4.3%	6.0%	15.7%	15.6%	54.8%	100.0%

Table 7.13

Holdings by Size of SHEEP BREEDING FLOCK
June Census 1991

	1 - <50	50 - <100	100 - <200	200 - <500	500 - <1000	1000 and over	Total
United Kingdom							
Number of holdings	26 042	16 108	18 114	19 399	7 762	3 031	90 456
percent	28.8%	17.8%	20.0%	21.4%	8.6%	3.4%	100.0%
Number of sheep	590 627	1 147 212	2 554 092	6 054 087	5 297 526	4 541 144	20 184 688
percent	2.9%	5.7%	12.7%	30.0%	26.2%	22.5%	100.0%
England							
Number of holdings	15 036	7 879	9 481	10 061	3 376	1 003	46 836
percent	32.1%	16.8%	20.2%	21.5%	7.2%	2.1%	100.0%
Number of sheep	316 801	564 671	1 343 286	3 129 029	2 280 664	1 472 995	9 107 446
percent	3.5%	6.2%	14.7%	34.4%	25.0%	16.2%	100.0%
Wales							
Number of holdings	3 629	2 555	3 051	4 601	2 398	921	17 155
percent	21.2%	14.9%	17.8%	26.8%	14.0%	5.4%	100.0%
Number of sheep	85 850	182 420	436 083	1 482 285	1 645 437	1 348 945	5 181 020
percent	1.7%	3.5%	8.4%	28.6%	31.8%	26.0%	100.0%
Scotland							
Number of holdings	3 355	2 892	3 189	3 323	1 782	1 065	15 606
percent	21.5%	18.5%	20.4%	21.3%	11.4%	6.8%	100.0%
Number of sheep	85 175	207 645	450 944	1 042 236	1 238 969	1 667 503	4 692 472
percent	1.8%	4.4%	9.6%	22.2%	26.4%	35.5%	100.0%
Northern Ireland							
Number of holdings	4 022	2 782	2 393	1 414	206	42	10 859
percent	37.0%	25.6%	22.0%	13.0%	1.9%	0.4%	100.0%
Number of sheep	102 801	192 476	323 779	400 537	132 456	51 701	1 203 750
percent	8.5%	16.0%	26.9%	33.3%	11.0%	4.3%	100.0%

Holdings by Total Number of SHEEP
June Census 1991

Table 7.14

	1 - <50	50 - <100	100 - <200	200 - <500	500 - <1000	1000 and over	Total
United Kingdom							
Number of holdings	17 457	11 617	15 622	22 620	14 407	12 256	93 979
percent	18.6%	12.4%	16.6%	24.1%	15.3%	13.0%	100.0%
Number of sheep	396 407	845 425	2 258 721	7 317 245	10 182 034	22 363 206	43 363 038
percent	0.9%	1.9%	5.2%	16.9%	23.5%	51.6%	100.0%
England							
Number of holdings	10 965	5 841	7 678	11 745	7 554	5 379	49 162
percent	22.3%	11.9%	15.6%	23.9%	15.4%	10.9%	100.0%
Number of sheep	234 456	423 911	1 111 977	3 816 332	5 318 197	9 344 399	20 249 272
percent	1.2%	2.1%	5.5%	18.8%	26.3%	46.1%	100.0%
Wales							
Number of holdings	2 246	1 815	2 502	4 022	3 392	3 536	17 513
percent	12.8%	10.4%	14.3%	23.0%	19.4%	20.2%	100.0%
Number of sheep	53 792	131 177	364 448	1 326 167	2 450 728	6 455 831	10 782 143
percent	0.5%	1.2%	3.4%	12.3%	22.7%	59.9%	100.0%
Scotland							
Number of holdings	1 862	1 785	2 786	3 956	2 503	3 062	15 954
percent	11.7%	11.2%	17.5%	24.8%	15.7%	19.2%	100.0%
Number of sheep	44 929	131 977	402 407	1 274 910	1 771 746	6 131 460	9 757 429
percent	0.5%	1.4%	4.1%	13.1%	18.2%	62.8%	100.0%
Northern Ireland							
Number of holdings	2 384	2 176	2 656	2 897	958	279	11 350
percent	21.0%	19.2%	23.4%	25.5%	8.4%	2.5%	100.0%
Number of sheep	63 230	158 360	379 889	899 836	641 363	431 516	2 574 194
percent	2.5%	6.2%	14.8%	35.0%	24.9%	16.8%	100.0%

Holdings by Size of LAYING FOWLS FLOCK
June Census 1991

Table 7.15

	1 - <25	25 - <50	50 - <100	100 - <200	200 - <1000	1000 - <5000	5000 - <20000	20000 and over	Total
United Kingdom									
Number of holdings	26 130	2 837	1 334	583	868	855	643	296	33 546
percent	77.9%	8.5%	4.0%	1.7%	2.6%	2.5%	1.9%	0.9%	100.0%
Number of birds	291 928	90 933	87 549	76 688	358 914	2 177 071	6 226 806	23 740 500	33 050 389
percent	0.9%	0.3%	0.3%	0.2%	1.1%	6.6%	18.8%	71.8%	100.0%
England									
Number of holdings	16 527	1 932	1 002	481	750	694	492	235	22 113
percent	74.7%	8.7%	4.5%	2.2%	3.4%	3.1%	2.2%	1.1%	100.0%
Number of birds	184 893	62 223	66 280	63 471	309 351	1 739 869	4 731 083	19 535 782	26 692 952
percent	0.7%	0.2%	0.2%	0.2%	1.2%	6.5%	17.7%	73.2%	100.0%
Wales									
Number of holdings	4 384	340	100	38	42	57	30	8	4 999
percent	87.7%	6.8%	2.0%	0.8%	0.8%	1.1%	0.6%	0.2%	100.0%
Number of birds	44 292	10 753	6 520	4 843	19 645	162 182	291 943	474 093	1 014 271
percent	4.4%	1.1%	0.6%	0.5%	1.9%	16.0%	28.8%	46.7%	100.0%
Scotland									
Number of holdings	3 476	386	139	38	67	51	39	17	4 213
percent	82.5%	9.2%	3.3%	0.9%	1.6%	1.2%	0.9%	0.4%	100.0%
Number of birds	39 510	12 245	9 376	5 225	27 358	122 049	379 300	1 999 929	2 594 992
percent	1.5%	0.5%	0.4%	0.2%	1.1%	4.7%	14.6%	77.1%	100.0%
Northern Ireland									
Number of holdings	1 743	179	93	26	9	53	82	36	2 221
percent	78.5%	8.1%	4.2%	1.2%	0.4%	2.4%	3.7%	1.6%	100.0%
Number of birds	23 233	5 712	5 373	3 149	2 560	152 971	824 480	1 730 696	2 748 174
percent	0.8%	0.2%	0.2%	0.1%	0.1%	5.6%	30.0%	63.0%	100.0%

Holdings by Size of TABLE FOWLS FLOCK

Table 7.16   June Census 1991

	1 - <1000	1000 - <20000	20000 - <100000	100000 and over	Total
United Kingdom					
Number of holdings	1 507	469	625	183	2 784
percent	54.1%	16.8%	22.4%	6.6%	100.0%
Number of birds	72 303	4 625 746	29 286 731	41 697 150	75 681 930
percent	0.1%	6.1%	38.7%	55.1%	100.0%
England					
Number of holdings	1 056	327	473	144	2 000
percent	52.8%	16.4%	23.7%	7.2%	100.0%
Number of birds	58 523	3 008 580	22 837 851	29 686 032	55 590 986
percent	0.1%	5.4%	41.1%	53.4%	100.0%
Wales					
Number of holdings	210	17	26	6	259
percent	81.1%	6.6%	10.0%	2.3%	100.0%
Number of birds	6 543	129 748	1 325 000	3 230 443	4 691 734
percent	0.1%	2.8%	28.2%	68.9%	100.0%
Scotland					
Number of holdings	148	27	25	26	226
percent	65.5%	11.9%	11.1%	11.5%	100.0%
Number of birds	3 697	273 638	1 284 680	7 495 375	9 057 390
percent	0.0%	3.0%	14.2%	82.8%	100.0%
Northern Ireland					
Number of holdings	93	98	101	7	299
percent	31.1%	32.8%	33.8%	2.3%	100.0%
Number of birds	3 540	1 213 780	3 839 200	1 285 300	6 341 820
percent	0.1%	19.1%	60.5%	20.3%	100.0%

Table 7.17

Holdings by FULL-TIME FAMILY and HIRED WORKERS
June Census 1991

	1	2	3	4	5 - <10	10 - <15	15 and over	Total
United Kingdom								
Number of holdings	30 634	12 346	4 951	2 334	3 253	1 194		54 712
percent	56.0%	22.6%	9.0%	4.3%	5.9%	2.2%		100.0%
Number of workers	30 634	24 692	14 853	9 336	20 172	25 862		125 549
percent	24.4%	19.7%	11.8%	7.4%	16.1%	20.6%		100.0%
England								
Number of holdings	20 063	8 625	3 594	1 835	2 724	540	519	37 900
percent	52.9%	22.8%	9.5%	4.8%	7.2%	1.4%	1.4%	100.0%
Number of workers	20 063	17 250	10 782	7 340	16 932	6 248	16 982	95 597
percent	21.0%	18.0%	11.3%	7.7%	17.7%	6.5%	17.8%	100.0%
Wales								
Number of holdings	2 733	748	205	62	50	18		3 816
percent	71.6%	19.6%	5.4%	1.6%	1.3%	0.5%		100.0%
Number of workers	2 733	1 496	615	248	300	338		5 730
percent	47.7%	26.1%	10.7%	4.3%	5.2%	5.9%		100.0%
Scotland								
Number of holdings	5 027	2 457	1 050	410	447	59	41	9 491
percent	53.0%	25.9%	11.1%	4.3%	4.7%	0.6%	0.4%	100.0%
Number of workers	5 027	4 914	3 150	1 640	2 734	679	1 195	19 339
percent	26.0%	25.4%	16.3%	8.5%	14.1%	3.5%	6.2%	100.0%
Northern Ireland								
Number of holdings	2 811	516	102	27	32	8	9	3 505
percent	80.2%	14.7%	2.9%	0.8%	0.9%	0.2%	0.3%	100.0%
Number of workers	2 811	1 032	306	108	206	98	322	4 883
percent	57.6%	21.1%	6.3%	2.2%	4.2%	2.0%	6.6%	100.0%

Notes

Data for England, Wales and Northern Ireland exclude farmers, partners, directors and their spouses together with salaried managers and trainees.

Data for Scotland differ in that they include salaried managers and partners and directors other than principal ones.

Table 7.18

Holdings by TENURE
June Census 1991

	Holdings owned or mainly owned		Holdings rented or mainly rented		Area owner occupied		Area rented		Total holdings	
	Number	%	Number	%	Hectares	%	Hectares	%	Number	Hectares
Great Britain										
Under 2 (ha.)	10 425	4.9	1 805	0.9	11 652	0.1	2 015	0.0	12 230	13 667
2 - < 20 (ha.)	58 836	27.8	13 860	6.6	527 698	3.3	137 035	0.9	72 696	664 733
20 - < 200 (ha.)	78 309	37.0	33 799	16.0	5 189 901	32.2	2 539 072	15.8	112 108	7 728 973
200 and over (ha.)	8 536	4.0	6 008	2.8	4 463 698	27.7	3 222 816	20.0	14 544	7 686 514
Total (ha.)	156 106	73.8	55 472	26.2	10 192 949	63.3	5 900 935	36.7	211 578	16 093 884
England										
Under 2 (ha.)	8 770	5.8	1 147	0.8	9 667	0.1	1 226	0.0	9 917	10 893
2 - < 20 (ha.)	44 608	29.5	8 834	5.9	394 294	4.2	91 639	1.0	53 442	485 933
20 - < 200 (ha.)	53 556	35.5	24 739	16.4	3 509 598	37.6	1 857 607	19.9	78 295	5 367 205
200 and over (ha.)	5 449	3.6	3 863	2.6	1 967 534	21.1	1 500 481	16.1	9 312	3 468 015
Total (ha.)	112 383	74.4	38 583	25.6	5 881 094	63.0	3 450 952	37.0	150 966	9 332 046
Wales										
Under 2 (ha.)	872	2.9	79	0.3	1 021	0.1	87	0.0	951	1 108
2 - < 20 (ha.)	9 276	31.2	1 591	5.4	89 628	6.0	17 283	1.2	10 867	106 911
20 - < 200 (ha.)	13 182	44.4	3 781	12.7	803 293	53.8	246 321	16.5	16 963	1 049 614
200 and over (ha.)	698	2.3	231	0.8	240 095	16.1	94 537	6.3	929	334 632
Total (ha.)	24 028	80.9	5 682	19.1	1 134 037	76.0	358 227	24.0	29 710	1 492 264
Scotland										
Under 2 (ha.)	783	2.5	579	1.9	964	0.0	702	0.0	1 362	1 666
2 - < 20 (ha.)	4 952	16.0	3 435	11.1	43 776	0.8	28 113	0.5	8 387	71 889
20 - < 200 (ha.)	11 571	37.4	5 279	17.1	877 010	16.6	435 144	8.3	16 850	1 312 154
200 and over (ha.)	2 389	7.7	1 914	6.2	2 256 069	42.8	1 627 798	30.9	4 303	3 883 867
Total (ha.)	19 695	63.7	11 207	36.3	3 177 818	60.3	2 091 756	39.7	30 902	5 269 574

Note

Totals may not necessarily agree with the sum of their components due to rounding.

Printed in the United Kingdom for HMSO.
Dd.0295569, 12/92, C10, 3397/5, 5673, 222275.

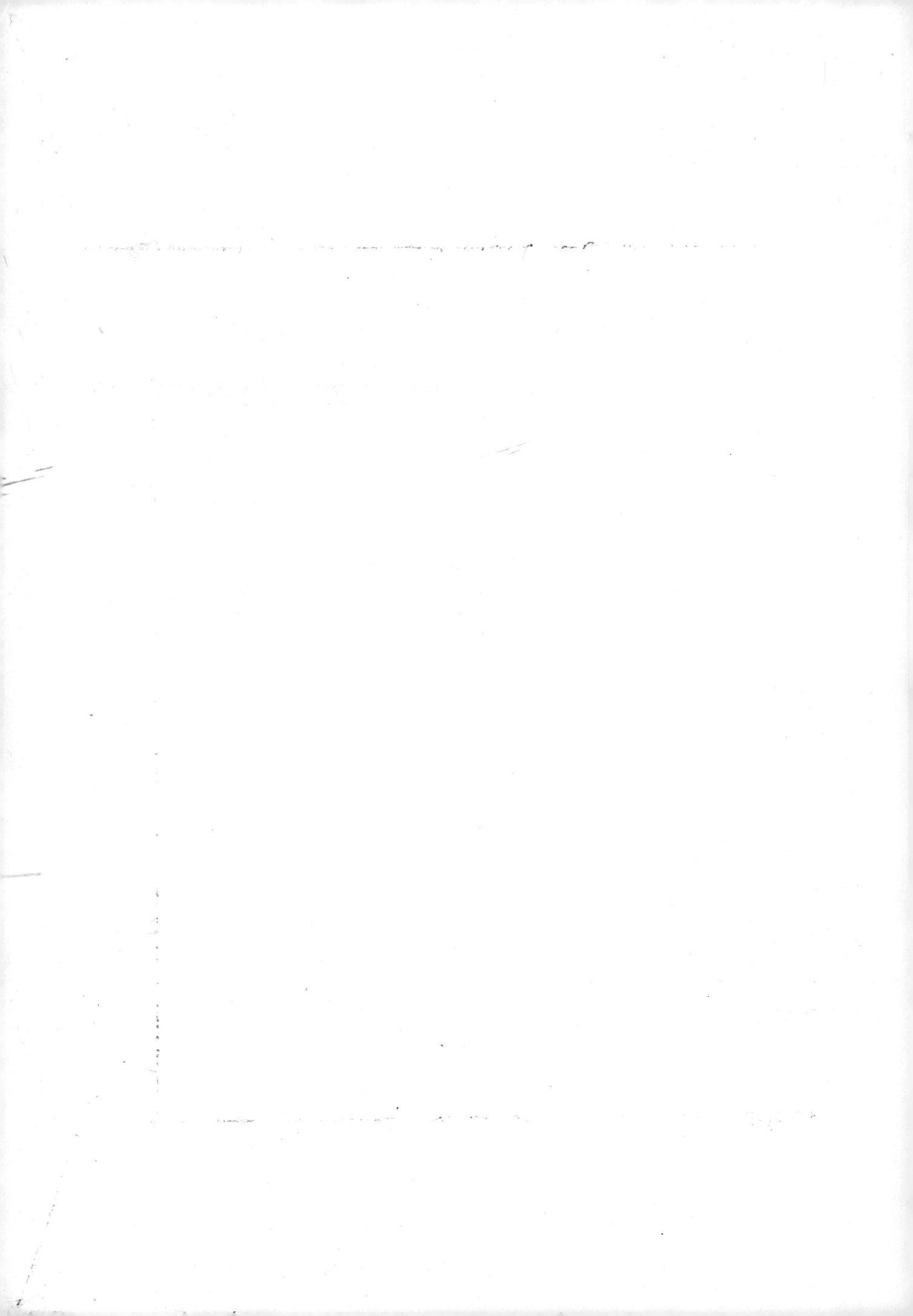